AF579293

THE TRUST CRISIS

Raj Ananthanpillai

THE TRUST CRISIS

How Big Tech Stole Your Identity—
and the New Model That **Takes It Back**

Forbes | Books

Published by Forbes Books, Charleston, South Carolina.
An imprint of Advantage Media Group.

Forbes Books is a registered trademark, and the Forbes Books colophon is a trademark of Forbes Media, LLC.

Printed in the United States of America.

10 9 8 7 6 5 4 3 2 1

ISBN: 979-8-88750-880-1 (Hardcover)
ISBN: 979-8-88750-881-8 (eBook)
ISBN: 979-8-88750-882-5 (Audiobook)

Library of Congress Control Number: 2026902681

Cover design by Ruthie Wood.
Layout design by Megan Elger.

Since 1917, Forbes has remained steadfast in its mission to serve as the defining voice of entrepreneurial capitalism. Forbes Books, launched in 2016 through a partnership with Advantage Media, furthers that aim by helping business and thought leaders bring their stories, passion, and knowledge to the forefront in custom books. Opinions expressed by Forbes Books authors are their own. To be considered for publication, please visit **books.Forbes.com**.

03-04-2026 2:13

To my beloved family, the quiet strength behind every step of this journey. To Radhika, my wife and life partner. To my children, Rekha and Raman, the light you bring into my world is brighter than any success I could ever claim.

And to my parents and my older and younger brothers who watch over me now from above, your love still lifts me, your pride still carries me forward, and your voices remain the clearest I hear when I wonder if I have done enough.

This book, like everything that matters, belongs to all of you.

CONTENTS

Part III: How Everything Can Change for the Better

ABOUT THE AUTHOR

Raj Ananthanpillai is a serial entrepreneur, investor, inventor, and leading expert on digital trust, identity, and privacy in an age of widespread data use and institutional decline. Raj is a strong supporter of individuals' data independence and ownership.

As founder and CEO of Trua, he is creating a new trust layer for the internet: reusable, privacy-focused digital credentials that allow individuals to securely prove their identity, qualifications, and background without losing control of their personal data. He has built and grown several impactful tech companies and founded Endera (which was acquired by Trua in 2023 and is now its risk intelligence division), InfoZen—a top risk management solutions firm acquired by ManTech International in 2017 (later acquired by the Carlyle Group in 2022)—and NetBalance (acquired in 2000). His ventures have provided mission-critical solutions for federal agencies, including key contributions to various national security programs such as TSA PreCheck, Fortune 500 companies, and global organizations.

Holding two master's degrees—one in engineering physics and one in electrical engineering—Raj is an inventor with multiple US patents in dynamic trust scoring, intelligent systems management, and privacy-preserving identity frameworks. He has authored two previous books on resilient global systems architectures and has advised governments, financial institutions, and corporate boards on security, risk, connectivity, and management.

Beyond his professional achievements, Raj's passion for education spans beyond his industry, including early childhood education on arts and science, American history, and civics. He serves on the board of the Wolf Trap Foundation for the Performing Arts, as an advisory board member for the University of Kansas School of Electrical Engineering and Computer Science, and as a cabinet member of the White House Historical Association's National Council on White House History.

ACKNOWLEDGMENTS

I extend my heartfelt gratitude to several key individuals whose guidance and support have been invaluable. Special recognition goes to Jeff Highman, CTO of Trua, for his technical expertise and innovative perspectives that enriched my thinking. Likewise, Ed Mathias and Charlie Walton, esteemed members of our industry advisory board, provided critical insights and encouragement that helped shape the book's core arguments.

My children also deserve immense credit; their enthusiastic nudges inspired me to embark on this writing journey, originally suggesting I chronicle my path to achieving the American dream as an immigrant. Though I ultimately chose a different focus, their prompting fueled my determination to produce a work that urges society to break free from decades of passive reliance on flawed, manipulated trust-verification systems—and instead embrace a new paradigm in which individuals reclaim their central role in fostering authentic trust.

I owe an enormous debt of gratitude to the outstanding Forbes Books publishing team for their remarkable speed and professionalism in bringing this book to market at an accelerated pace that far exceeded my expectations. The entire team—spanning editorial, graphics, production, and marketing—worked tirelessly and collaboratively to refine every detail, polish the manuscript, and ensure a flawless, timely launch. Their dedication and efficiency were truly instrumental in transforming my vision into reality so swiftly.

Finally, I must express my profound thanks to my wife, whose unwavering patience during my countless late evening hours at the desk allowed this project to come to fruition.

PREFACE

In an era when our lives are inextricably woven into the fabric of digital and AI-driven ecosystems, trust has become a fragile commodity, eroded by relentless waves of data breaches, identity theft, and AI-driven deceptions.

This is the reason I founded Trua—to build the very alternative I had been searching for: a reusable, user-controlled trust credential that lets individuals get their identity screened and verified once, store it securely in their own digital wallet, and share only what is needed, without repeatedly giving out or ever exposing raw personally identifiable information (PII) such as Social Security number, date of birth, and more.

Trua is the practical embodiment of this book's vision. It's a living solution designed to end the needless proliferation of sensitive data, dramatically reduce fraud, and hand power back to the individual. Motivated by the personal stories of those affected by data breaches and the systemic flaws I had observed firsthand, I launched Trua to create a platform that leverages blockchain, biometrics, and zero-knowledge proofs to empower users with sovereignty over their information, making verification efficient for businesses while protecting privacy. It is not just about identifying problems, but about deploying real-world solutions to fix them, starting with pilot programs that have already shown reductions in verification times and fraud risks in sectors such as employment and the gig economy.

My journey through these pages is informed by years of writing, including *Forbes* articles on reusable credentials and democratizing screening, as well as the hands-on work of building Trua from the ground up. Whether you are a consumer weary of handing over your Social Security number yet again, a gig worker exhausted by repetitive background checks, or a business leader facing mounting compliance costs, reputational risks, or the threat of a breach (or you've experienced one), this book and the platform born from it equip you with both the understanding and the tools to reclaim control and deploy a trust layer for all internet commerce and interactions that benefits the new AI-driven digital society.

As AI propels us toward an interconnected future of autonomous systems and emerging threats, the time to act is now. Trust is not given; it is built, verified, and protected. Join me in this movement for a world where sovereignty over our data restores not just security but the very foundations of our shared humanity, bound by trust.

INTRODUCTION

I still remember the afternoon in September 2017 when news broke that Equifax had been hacked. As one of the three major credit bureaus, Equifax held sensitive personal data on hundreds of millions of people, and now the information of 147 million people had been suddenly exposed to unknown criminals.

There was a sinking feeling in my stomach, not just as an industry veteran but as a person who had just become a statistic of having his data compromised. In my decades working in digital identity and security, I had always known the system was fragile. But the Equifax breach was different. It underscored, in the most glaring way, that our entire system of trust is and has been fundamentally broken.

The institutions supposedly safeguarding our data had failed us, and we suffered the consequences while they walked away with minor scratches. Equifax wound up paying over half a billion dollars in fines and settlements,[1] a hefty sum. Yet the company's stock bounced back, and its business continues to grow. Meanwhile, tens of millions of

1 "Equifax to Pay $575 Million as Part of Settlement with FTC, CFPB, and States Related to 2017 Data Breach," press release, Federal Trade Commission, July 22, 2019, https://www.ftc.gov/news-events/news/press-releases/2019/07/equifax-pay-575-million-part-settlement-ftc-cfpb-states-related-2017-data-breach.

ordinary people like you and me were left to freeze our credit, watch our bank accounts, and worry for years on end about identity theft.

Moments like that carry a heavy emotional weight for me. They fuel my resolve that something has to change. How did we get to a world where we routinely hand over the keys to our identity and hope that faceless corporations will guard them? Why do we continue to trust a broken system with our most valuable personal information?

I come to this problem, and to this book, with a very personal stake and a deep well of experience. Over my career, I've been on the front lines of digital identity and trust. I've led a company that developed the TSA PreCheck program for the government, speeding up airport security by verifying trust up front.

I've advised businesses on cybersecurity and watched as data breaches exploded into a constant threat. I even witnessed from the inside how a major federal background check contractor, USIS, fell victim to hackers in 2014—an attack that exposed the personal files of at least twenty-five thousand Department of Homeland Security employees[2] and ultimately caused USIS to lose its government contracts, leading to the shutdown of its operations.

I've sat in war rooms during breach responses, hearing the anxiety in executives' voices as they realize they've lost their customers' trust overnight. And I've spoken to countless friends, colleagues, and everyday people about the personal toll of these failures. I've heard stories of credit scores ruined by identity theft, of job offers lost due to erroneous background check data, and of creeping fear each time another breach headline pops up. These experiences have seared into

2 David Bisson, "The OPM Breach: Timeline of a Hack," Fortra, last updated July 10, 2015, https://www.tripwire.com/state-of-security/the-opm-breach-timeline-of-a-hack.

me the simple truth that the way we build digital trust and verify today is not working. In fact, it's hurting us.

We live in an era when trust has been eroded by rampant data abuse, breaches, and a feeling of powerlessness. By mid-2025, 26 percent of US consumers had been affected by identity theft in the prior two years, up sharply from 18 percent just a couple of years before. Each victim of identity theft loses about $1,000 on average, not to mention the immeasurable stress. And it's not just a US problem. Globally, one in four adults was hit by some form of digital fraud in 2024, with many suffering financial harm, and around 67 percent of everyday victims—your customers, employees, neighbors—say they've contemplated self-harm because of identity crime.[3]

> *The way we build digital trust and verify today is not working.*

These statistics represent the trauma of a single parent drained of savings by a scam, the young graduate whose future is derailed by a stolen identity, and the elderly couple afraid to shop online anymore. The emotional and psychological toll of living in a low-trust digital world is profound, and we've been let down by the very institutions meant to protect us.

A Crisis of Trust

This loss of trust spreads like cracks in a foundation, threatening the entire digital economy we depend on. When people don't trust, they pull back. According to Pew research, "The public increasingly says they don't understand what companies are doing with their data. Some

3 *ITRC 2025 Consumer Impact Report* (Identity Theft Resource Center, 2025), https://www.idtheftcenter.org/publication/itrc-2025-consumer-impact-report/.

67% say they understand little to nothing about what companies are doing with their personal data."[4] And over half of Americans have stopped using a product or service because of privacy concerns.[5]

I've spoken to business leaders who see this in their own customer bases, with users opting out, withholding information, or demanding deletion of their accounts. Lost trust is devastating for business.

Research consistently shows that data breaches destroy customer trust and loyalty. A large majority of consumers say they would stop buying from a company after a cybersecurity incident, with 75 percent of US consumers reporting they would abandon a brand following a breach and 66 percent saying they would no longer trust that company with their data.[6] This is a full-blown crisis of confidence. Trust isn't just a nice-to-have. It's become a bottom-line, societal, and personal issue for each of us.

Stuck in the Past

In the early days of the internet (I'm talking the 1970s ARPANET era), there were essentially no security checks. The pioneers operated on unbridled optimism and trust; the community was small, and everyone pretty much knew each other. There was no concept of a digital identity, let alone verification, because it wasn't needed.

---

4 Colleen McClain et al., "How Americans View Data Privacy," *Pew Research Center*, October 18, 2023, https://www.pewresearch.org/internet/2023/10/18/how-americans-view-data-privacy/.

5 Andrew Perrin, "Half of Americans Have Decided Not to Use a Product or Service Because of Privacy Concerns," *Pew Research Center*, April 14, 2020, https://www.pewresearch.org/short-reads/2020/04/14/half-of-americans-have-decided-not-to-use-a-product-or-service-because-of-privacy-concerns/.

6 Dana Kringel, "Vercara Research: 75% of U.S. Consumers Would Stop Purchasing from a Brand if It Suffered a Cyber Incident," Vercara, December 18, 2023, https://vercara.digicert.com/news/vercara-research-75-of-u-s-consumers-would-stop-purchasing-from-a-brand-if-it-suffered-a-cyber-incident?utm_source=chatgpt.com.

Fast forward to today, and those naïve early designs have given way to a Wild West where everything about us is online yet often unverified. Over time, we moved from "trust by default" to "trust but verify" and now to "verify first, then trust" as the standard approach. Why? Because bad actors flooded the system.

When the internet opened to the masses, criminals and scammers weren't far behind, exploiting anonymity and weak protections. By the 2010s, massive data breaches had become commonplace. Names such as Yahoo, Target, the Office of Personnel Management, and Marriott became synonymous with personal data leaks of tens or hundreds of millions of individuals. We've learned through painful experience that on the internet, you verify before you trust.

And yet, despite this hard-earned wisdom, our basic model for verification and trust remains stuck in the past. That's what this book is here to change. Think about what you do today whenever you start a new job, create a new account, or use a new service. You hand over your personal information repeatedly: name, address, Social Security number (SSN), ID, biometric scans—an endless parade of personally identifiable information (PII) given out to each entity that asks.

When applying for an apartment, you give your SSN to the landlord for a credit check. When signing up for a dating app, you typically share your Facebook profile or phone number. Want a store loyalty card? Sure, give them your email, phone number, and maybe your birthday too.

We have been conditioned to surrender or freely give out our personal data without a second thought, thanks in part to social media and other digital platforms. Our complacency and eagerness for convenience have created a new golden age of piracy, a world where there is an ocean of personal data out there for hackers to pillage.

Each piece of data we give away sits in yet another database, forming an ever-expanding attack surface. Your data isn't just in one place; it's scattered across hundreds of databases like tempting honeypots for cybercriminals. As a result, even routine daily activities such as ordering groceries, logging in to social media, and applying for a job carry an undercurrent of risk that simply didn't exist a generation ago.

The Cost of Complacency

Our complacency in clicking "Accept" has been interpreted as blanket approval for these practices. It infuriates me, and it should infuriate you too. We've been lulled into thinking this is just the way it is. *Of course apps gather data, of course we're going to get targeted ads, and of course companies need our information to serve us.*

But do they really? Do grocery stores really need your phone number and birthday just to give you a discount? Does a random online retailer really need to store your credit card and address forever?

Often, it's done simply because everyone else does it, and because personal data has become an enormously valuable asset on the corporate balance sheet. Companies today default to hoarding data because somewhere along the line, they forgot the why and only saw the dollar signs. Yet by stockpiling all this sensitive information, they've unintentionally created massive liability for themselves with treasure troves that attract hackers like flies to honey. They collect all this personal data supposedly to better secure and serve their customers, but in doing so, they've created the very vulnerability that now plagues us.

The result of these perverse incentives is a vicious cycle of data proliferation and breach risk. The more places your data lives, the more chances for it to leak. And when it leaks, it can be used against you in countless ways, such as identity theft, financial fraud, phishing scams, even personal extortion. Twenty-six percent of US consumers have been affected by identity theft in the past two years alone, up from 18 percent in 2023, with losses totaling over $10 billion annually.[7]

Cybercriminals find PII incredibly valuable. When armed with sensitive data such as SSNs, criminals can open fake credit lines, file false tax returns, or create convincing impersonation scams. And guess who foots the bill? You do, either directly or through higher costs passed down by businesses and insurers.

By 2024, the global cost of data breaches had soared to $46.25 billion annually. The average breach in the US costs a company about $4.9 million, but the downstream costs to individuals and society are even greater.[8] We're talking hours of lost productivity resolving identity fraud cases, the psychological impact of privacy violations, erosion of trust in commerce, and so on.

There's also an insidious chilling effect as people become less willing to engage in beneficial activities such as e-commerce, online learning, and digital health services because of fear that their data will be misused. A 2025 Edelman Trust Barometer report noted that 68 percent of global consumers avoid brands that they perceive as lax on data security.[9] So companies that should be innovating and offering great services are instead spending resources on damage control and expensive cybersecurity insurance. It's a high-cost, low-trust equilibrium, and nobody truly wins except maybe the cybercrime rings.

---

7 Doug Bonderud, "Cost of a Data Breach 2024: Financial Industry," IBM, accessed December 17, 2025, https://www.ibm.com/think/insights/cost-of-a-data-breach-2024-financial-industry.

8 Ibid.

9 *2025 Edelman Trust Barometer Global Report: Trust and the Crisis of Grievance* (Edelman Trust Institute), January 17, 2025, https://www.edelman.com/sites/g/files/aatuss191/files/2025-01/2025%20Edelman%20Trust%20Barometer_Final.pdf.

What gets me especially fired up is that individuals currently have little to no agency in this whole picture. We've been powerless by design. Sure, there are laws such as the California Consumer Privacy Act or Europe's General Data Protection Regulation that grant certain rights, including the right to request your data and delete it, but in practice, exercising those rights is tedious and after-the-fact. Besides, some of those laws are one-sided and end up hampering business in terms of verifying consumers before providing services or hiring them for employment. You usually find out your data was abused long after the damage is done. In the moment of decision, the moment you sign up for that new service or job, you typically get a binary choice: Hand over your data or walk away.

It's all or nothing, and realistically, saying no means you miss out on the service. As I discussed with colleagues, many in the younger generation now just shrug and say of course we must give up data to get something, because they've never seen an alternative. But deep down, I know, and surveys confirm, that people want more control.

A Better Way

If you're a consumer, it's critical that you wake up and realize the severity of the problem. Don't wait until a crisis emerges before you act. Take proactive steps now. If you're a business owner, frankly, you need to remember your moral obligation to those you serve and be part of redesigning the system from the ground up. Stop being part of the problem: Give your users clear control of their privacy settings and stop hiding critical details behind endless pages of three-point-font text and legal jargon.

In the interplay between corporate lobbyists and regulatory authorities—entities ostensibly tasked with safeguarding consumer interests—a tacit collaboration frequently emerges. Lengthy legal disclaimers, printed in minuscule font, serve primarily as formalities to fulfill compliance requirements and expedite approval. Rarely do consumers engage with or read these documents, rendering genuine protection secondary to the ritual of regulatory adherence. In other words, it becomes merely a check-the-box exercise.

We all need to work together to flip the model so that individuals like you own and control their personal data and share it on their terms, rather than dozens of third parties collecting and controlling it on your behalf. We need to enable trust without constant data duplication. We need a way for you to prove things about yourself—that you are who you claim, that you have a clean record, that you're qualified for a job or trustworthy to rent an apartment—*without* handing over all your underlying personal details every single time.

In this book, I will argue that we need a "Trust Credential for Life" and a supportive Trust Bureau to make it work. The Trust Credential is essentially your digital passport for all identity and background needs. Instead of handing over your Social Security number, driver's license, passport, and a file of background check documents to every entity that asks, you would have one reusable digital credential issued after a thorough verification of your identity and records that you have agency over it.

That is the ultimate win-win solution I offer for both consumers and businesses, but before we get there, it's important to understand why we are in the mess we're in today and the depth of the challenges we face.

PART I

THE NEED FOR YOUR OWN REUSABLE TRUST CREDENTIAL

CHAPTER 1

HOW THE SYSTEM WAS ALWAYS RIGGED AGAINST YOU

If you feel like the digital world has been taking advantage of you, you're not wrong. In fact, it was built that way. Our current digital trust system, if we can even call it trust, was rigged against individuals from the very start.

This might sound like a conspiracy, but it's really a story of misaligned incentives and historical baggage. In this chapter, I want to pull back the curtain on how we arrived at a system where your personal data fuels other people's profits and your security is an afterthought. By the end of this, you'll see clearly that the playing field was never level and that we desperately need a new one.

To understand the rigged game, let's start with one of the oldest players: the credit bureaus. They began over a century ago as simple credit reporting agencies but, over time, became the unaccountable (albeit regulated) gatekeepers of your financial creditworthiness for banks, lenders, and purchases you want to make. The idea, initially, was sensible: Centralize and standardize information on who is creditworthy so lenders can make informed decisions. And indeed, credit bureaus such as Equifax and Experian revolutionized financial trust after the passing of regulatory changes, including the Fair Credit Reporting Act in 1970, by aggregating data and providing a single credit report and score that banks rely on.

Before the advent of credit bureaus compiling reports and issuing FICO scores, lenders painstakingly crafted their own due diligence. This was often inconsistent, which bred inefficiencies and risks for financial institutions. Credit bureaus introduced a standardized report and score, readily interpretable by any relying party, enabling confident provision of goods and services without constant second-guessing.

The flipside, however, was that we consumers lost control over our own financial reputations. Today, credit bureaus hold files on literally billions of individuals globally, and none of us gave explicit consent. You don't sign up for a credit bureau. They just scoop up your data from loans, credit cards, mortgages, court records, and more. They then turn around and sell that data in the form of credit reports, scores, and marketing lists.

Lenders pay them, landlords pay them, and employers pay them. We—the subjects of all this data—are not the *customers*; we're the *product*. Our personal financial behaviors, from whether we pay the electric bill on time to how much we owe on a car loan, are packaged and monetized by agencies that operate in the background of our lives.

What's worse, the credit bureaus historically haven't been too concerned with accuracy or fairness, because their clients are the creditors and businesses, not us. Studies over the years have found alarming error rates in credit reports. A mistake in your file can wreck your ability to get a loan or a job. Yet trying to fix those errors is an exercise in frustration. You have to go through a formal dispute process, wait weeks or months, and hope the bureau or the creditor agrees.

Meanwhile, the burden is on *you* to prove the report wrong, rather than on them to get it right. It's utterly backward. The credit bureaus also epitomize the "collect everything, keep it forever" approach to data. They have records going back decades and will keep a paid-off loan or a closed credit card in your report for seven to ten years.

From their perspective, more data is always better, as it increases the value of their product. From your perspective, it's a privacy and security time bomb. The longer data sits around, the more likely it is to be breached or misused. And indeed, in 2017, that bomb exploded when Equifax failed to patch a known vulnerability, and hackers made off with the personal information of nearly half the US population. Consumers are suffering the consequences of that massive breach to this day.

The people whose data was stolen had zero say in Equifax holding it in the first place. Yet they were the ones who paid the price, through potential identity theft and the hassle of credit monitoring. Equifax's business, after some public shaming and payouts, continues on, but regular individuals cannot simply opt out of the credit bureau system. You're in, whether you like it or not. This is Exhibit A of how the system is rigged. An institution holds power over your financial identity without your consent or control.

The trust system was also rigged in another way through information asymmetry. Credit bureaus and similar entities operate opaquely. You often don't know when someone checks your data (whether part of it or all of it), and you certainly aren't asked each time. It's a one-way mirror where businesses see in, but you can't see out. Only when something goes wrong (such as you're denied a loan) do you even realize there might be an issue. And if a breach happens, you're typically notified long after the fact. For the most part, we've been kept in the dark regarding who accesses our information and for what purpose.

In a very real sense, you don't truly own "your" data under this model; the data brokers do. They'll argue that they're just stewards or intermediaries, but the control dynamics say otherwise. If you can't prevent collection, can't readily correct errors, can't stop them from selling it, and can't easily walk away, then it's not your data anymore. Credit bureaus proved that centralized verification can be efficient by streamlining lending decisions and likely added trillions in economic growth by extending credit more broadly. But they also showed that when people lack agency over their information, they are inevitably exploited.

The *concept* of a bureau is not inherently evil; it's the *execution* we ended up with that's exploitative. This is why, as we'll discuss later, my idea of a Trust Bureau consciously inverts many elements of the credit bureau model. The old model set the rules to favor institutions; the new model must reset them to empower individuals while servicing business needs.

The old model set the rules to favor institutions; the new model must reset them to empower individuals while servicing business needs.

The Illusion of Choice

Now let's talk about manufactured consent, one of the craftiest tricks in the digital playbook. Every time you install an app, make an account, or even use a Wi-Fi hotspot, you are presented with a terms-of-service or privacy policy and typically a box to check: "I agree."

Let's be honest. None of us read them in full. If we did, we'd spend half our lives poring over legal jargon. Companies know this, and those documents are designed less to inform you and more to protect the company and comply with regulatory requirements. Buried inside are clauses that often allow nearly boundless use of your data. They might say they can share your data with "affiliates" or "third-party partners," which can be a huge array of advertisers and data brokers.

They'll note that you grant them license to use any content you post, so that cute photo you uploaded can be repurposed in a marketing campaign without further permission. And of course, they'll disclaim liability for just about anything that goes wrong. By clicking "I agree," you're effectively signing a contract that overwhelmingly favors the other side.

This is consent theater. It gives the impression that you have a choice, but in practice, you don't. If you need the app or service, you will click agree, because it's take it or leave it. There's no negotiation on terms. Sure, in theory, you could refuse and forgo the service, but in today's world, that can mean excluding yourself from essentials.

Companies have manufactured our consent at scale, such that when privacy advocates say, "This is outrageous," companies can retort: "Well, users agreed to it." Did we, though? Is it *meaningful consent* when the alternative is opting out of modern life? I argue it is not. It's *coercive consent*, and it's a cornerstone of how the system stays rigged.

The legal system, until recently, offered scant help because courts largely treat those click-through agreements as binding contracts, even though everyone knows they aren't read. Thankfully, public pressure is forcing some change through demanding simpler disclosures or specific opt-outs for certain data uses, but we have a long way to go.

From a personal vantage point, I find this part of the system dehumanizing. I've heard of tech companies where certain decisions are made about data collection. Often, the discussion is around what they can get away with rather than what's best for users. "Can we use location data for targeted ads? Sure, put it in the privacy policy; nobody will object."

HOW THE SYSTEM WAS ALWAYS RIGGED AGAINST YOU

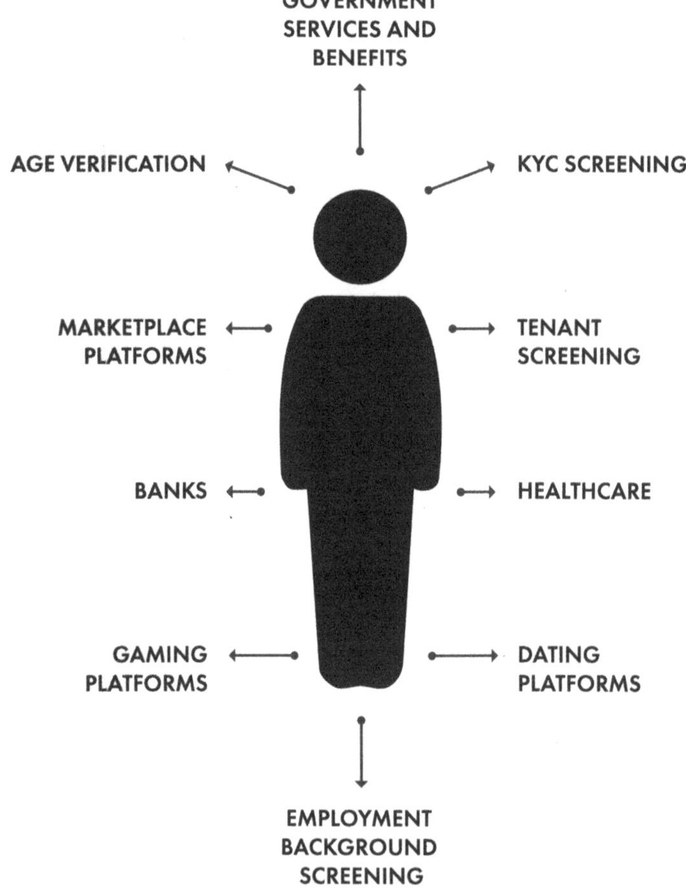

This cynicism pervades the industry. It's not usually ill-intentioned individuals; it's just the normalized culture. Everyone else does it, and there's money to be made, so why not? And if users complain, well, they clicked agree. It's as if a surgeon got you to sign a form allowing any procedure whatsoever, then did something unnecessary but profitable, and when confronted, points to your signature on page 19 of the fine print. You'd be livid, and rightfully so. Yet in digital services, that scenario plays out in some form every day.

Here's a striking example of how far this can go. On social media platforms, third-party apps historically could vacuum up not just your data but sometimes your friends' data too, all because somewhere in the labyrinthine settings, you and your friends "consented" to it. This is how Cambridge Analytica harvested data on tens of millions of Facebook users. A few hundred thousand people installed a quiz app and agreed to its terms, which then allowed it to pull the profiles of all their Facebook friends, building a massive dataset without those millions of people even realizing their data had been given away.[10]

It felt like a betrayal when this came to light. People said, "I never agreed to that!" only to learn that Facebook's policies at the time allowed it implicitly. The outrage was palpable. It was a moment that woke a lot of people up to the sham of digital consent. But even after Cambridge Analytica and similar scandals, not a ton changed structurally. Companies tightened a few policies and made privacy settings a little visible, but the fundamental imbalance remains.

Let's put some numbers on this imbalance. A *Forbes* survey in late 2023 found that 86 percent of Americans are more concerned about their privacy and data security than even the state of the economy, reflecting how front-of-mind this issue has become.[11] And yet, that

10 Adam Sanger, "Facebook Users' Data Harvested by Cambridge Analytica, Prompting U.S. Inquiries," *The New York Times*, April 8, 2018, https://www.nytimes.com/2018/04/08/us/facebook-users-data-harvested-cambridge-analytica.html.

11 Gary Drenik, "Data Privacy Tops Concerns for Americans—Who Is Responsible for Better Data Protections?," *Forbes*, December 8, 2023, https://www.forbes.com/sites/garydrenik/2023/12/08/data-privacy-tops-concerns-for-americans--who-is-responsible-for-better-data-protections/.

same study noted that about two-thirds of Americans feel resigned. They don't trust how their data is handled, and they aren't confident things will change.

This is the complacency and resignation I want to break. Because if we collectively demand change, companies will have to adjust. In 2024, several Big Tech firms scrambled to roll out new privacy options because users and regulators started pushing back. Apple introduced pop-ups asking you if you want to allow apps to track you across other apps. And guess what, when given a real choice, the vast majority of users say, "No, do not track me."

If we collectively demand change, companies will have to adjust.

That one move by Apple, empowering consumers with a clear yes/no choice, wiped out billions in targeted ad revenue for those apps, which tells you how invasive the tracking was behind the scenes. If more such power shifts happen, the rigged game starts to crumble.

You Are the Product

Another pillar of this rigged system is the concept that your personal data equals big money for others. It's often said that personal data is the new oil fueling the digital economy. I'd tweak that metaphor, because at least oil is something companies pay to extract. Personal data is more like groundwater. It just seeps out everywhere, and companies stake claims to pump it freely. We, the sources of that data, rarely get any compensation or even recognition.

Let's talk about the data broker industry. Most of you probably haven't interacted with a data broker directly, but they have definitely interacted with you or with your data. These are firms whose entire business is trading in personal information: credit card transactions, purchase histories, web browsing behaviors, location trails from your phone, demographic details, you name it.

They ingest data from various sources—some public, some commercial partnerships—and then they aggregate, slice, and dice it to sell as insights or lists. Ever wonder why, after browsing for a certain pair of shoes, you suddenly see shoe ads on every site? It's data brokers at work. The shoe retailer shared your interest via trackers, and a broker identified you across multiple contexts and sold your "shoe shopper" profile to advertisers in real time.

It's a sophisticated operation, largely invisible to users. And it's a multibillion-dollar industry. One leading data broker boasted of having profiles on over five hundred million consumers worldwide, each with thousands of data points. They cluster people into categories such as "expectant parents," "diabetes sufferers," or "luxury car likely buyers" and sell access to those lists. The people on the lists have no idea this is happening.

Unlike credit reports, which you can request and view, marketing data profiles are completely opaque. You typically can't see what companies have compiled about you, nor correct it if it's wrong. This can obviously lead to harm. Imagine data saying you have a certain medical condition when you don't, potentially affecting what health offers or insurance you see. Or data incorrectly pegging you as a "financial risk" so you're quietly blacklisted from certain offers.

The brokers will argue it's all fair game because the data is "anonymized" or "aggregated," but time and again, researchers have shown how trivial it is to reidentify individuals from supposedly anonymized datasets by cross-referencing a few unique data points. Plus, many brokers operate in a grey area where they actually have personally identifiable information (PII) tied to the profiles. They just don't always publicize that.

Where do they get this data? Some from publicly available records, such as property deeds, which include names and addresses. Some from purchasing it off companies that you interacted with. Some from web-tracking cookies and pixels that essentially spy on your online activity. We have built a sprawling ecosystem of surveillance capitalism where our lives are monitored, recorded, and monetized in exchange for "free" services or minor conveniences.

We have built a sprawling ecosystem of surveillance capitalism where our lives are monitored, recorded, and monetized in exchange for "free" services or minor conveniences.

It's worth highlighting social media again here. Social platforms encouraged us to share the most intimate details of our lives—what we like, whom we know, our milestones, our daily thoughts—under the promise of connecting with others. And connect we did. But behind the scenes, every like, every click, every dwell time on a post was tracked to build a profile for ad targeting.

Facebook's revenue soared into the tens of billions per quarter primarily because of how well it could target ads using our personal data. I don't begrudge a company for being successful, but the lack of transparency and control made it a devil's bargain. Users weren't meaningfully told, "Hey, by the way, we're going to infer your political leanings, your relationship status, and your impulsiveness from your activity and sell that insight to advertisers." That was just … done.

When it leaked that Facebook had categories such as "insecure about finances" or "interested in homosexuality" attached to users for ad targeting, people felt violated. Facebook didn't assign those labels maliciously; an algorithm did it based on behavior patterns. But to the person being microtargeted, it is creepy to be manipulated in that way without consent. Even national politicians have utilized such microtargeting techniques for political messaging and gain. The same goes for Google, which probably knows more about you than your own family does. Google uses that to target ads brilliantly and rake in revenue. Yet if you wanted to stop Google from profiling you, your only real option has been to stop using Google services—again, an almost impossible choice for most. Just for context, over three-quarters of Google's revenue is derived from online advertising,[12] which is what other companies pay Google to advertise based on your preferences, keywords in your Gmail, and your profiles on the internet.

So, we have a situation where companies get rich from personal data while individuals take on all the risk. Think about a bank. If they hold your money, they don't get to gamble with it scot-free. Regulations ensure they are custodians of your money and have responsibilities. If the bank fails, there's FDIC insurance, so you're protected. But in data, if a company holds your info and loses it, yes, they get a fine or bad press, but you may deal with the fallout for years, and there's no automatic compensation or insurance that truly covers the personal havoc. In a way, our personal data has been extracted like a natural resource, with minimal benefit flowing back to us and lots of externalities pushed onto us.

Our personal data has been extracted like a natural resource, with minimal benefit flowing back to us and lots of externalities pushed onto us.

12 "Google Ad Revenue (2013–2027)," Oberlo, accessed December 17, 2025, https://www.oberlo.com/statistics/google-ad-revenue.

This model inherently puts individuals at a disadvantage. Even if breaches were rare and data wasn't misused, the sheer asymmetry of knowledge and power is problematic. It's like a casino where the house always knows your hand, but you don't know theirs. Chances are you wouldn't want to play. Yet online, we play it every day. The house (Big Tech and data brokers) sees and analyzes our every move. We get only the information they decide to show us, often filtered and "personalized" in ways we can't discern.

I often think about the concept of manufactured profiles; not just what we knowingly provide but what the system infers about us. For example, you might never tell a shopping site your income level, but based on your browsing and buying patterns, it might algorithmically decide whether to show you high-end or budget products, effectively social-sorting you without your awareness.

People have reported seeing higher prices for the same item on an online store when using a Mac (perceived as wealthier) versus a PC. That's a crude example of how your data is used in ways that affect you materially. It's often subtle and behind the scenes, but it reinforces advantages for those who know how to game the system and leaves others paying more or missing out.

Breaches Everywhere

Our personal data is scattered all over the place. It's as if pieces of your identity are strewn across hundreds of digital vaults, and each vault has a door that can be picked by a determined thief. This proliferation of PII is at the core of why breaches have become so devastating. Remember that earlier stat: more than eight billion records exposed in 2023. That's not because one or two big companies got hacked; it's because practically every organization now holds a cache of personal data, and many of them will eventually suffer a breach. It's a numbers game.

Let's enumerate some types of breaches to appreciate the scope of the issue:

- Financial breaches: Banks and credit card companies are prime targets, obviously, but they actually invest heavily in security, so attackers often go after weaker links. They target that small e-commerce site where you saved your credit card once, or the payment processor behind your local store. Those get breached, and credit card numbers along with names/addresses get out. In 2024, we saw an increase in attacks on financial third parties, yielding millions of card numbers at a time.
- Healthcare breaches: Hospitals, clinics, and insurance companies all hold extremely sensitive info such as medical histories, prescriptions, and Social Security numbers (SSNs). The irony is that these institutions need that data to treat you or for legal compliance, but when breached, the info can be used for identity theft or even blackmail. In 2023, there was a major hospital system breach that exposed the data of eleven million patients; among the data were diagnoses and treatments.[13] Now, imagine that floating on the dark web. It's horrific.
- Government breaches: There have been incidents such as the US Office of Personnel Management (OPM) hack in 2015, where twenty-two million federal employees' records (including security clearance background check info with incredibly detailed personal data) were stolen in a breach believed to have been carried out by a nation-state.[14] That

13 Ibid.

14 Bisson, "The OPM Breach."

included fingerprints and personal narratives. Talk about being rigged against the individual. Those folks couldn't opt out of giving that info because it was required for their jobs. The government or its designated contractor company failed to secure it, and now hostile actors have it. There was also the USIS contractor breach mentioned previously, which was a prelude to OPM's big hack.[15] Foreign spies literally used our data overcollection against us.

- Education breaches: Universities hold student records, including financial and often health information. Many have been breached. In one case, a breach exposed thousands of students' identities and even their disciplinary records.
- Retail and hospitality breaches: Think of how many stores and hotels have your data. Marriott had a breach of five hundred million guest records in 2018. Target had a notorious breach of forty million credit cards in 2013. Every time you swipe a card or book a room, you're giving out data that might leak.

It's endless, and it's exhausting for individuals. We get those letters or emails: "We regret to inform you that your data may have been part of an incident. Here's free credit monitoring for a year." At this point, I've lost count of how many such notices I've personally received.

A national survey conducted by Talker Research (and sponsored by Trua) found that a significant portion of people just assume their data has been leaked at some point and are almost numb to it.[16] But being numb is dangerous because complacency sets in, which makes

15 Ibid.

16 Trua, *The State of Trust and Safety in Online Marketplaces*, February 11, 2025, accessed November 17, 2025, https://truame.com/trua-customer-survey-ebook/.

us click on phishing links more easily, share personal information freely, and perpetuate this harmful cycle.

When your data is out there, criminals can open credit lines, file false tax returns, or create fake driver's licenses in your name. For each victim, it can take months or years to unravel the damage, including dealing with creditors, restoring credit scores, explaining to banks, and maybe even clearing their name if crimes were committed using their identity.

I've helped some folks navigate identity theft, and it's a nightmare. One gentleman had someone take out an auto loan in his name, and by the time he discovered it, it had gone unpaid for months, and his credit was shot. He spent over a hundred hours trying to fix it, all because his SSN had leaked in a breach of a company he'd never even heard of. This illustrates the unfairness and how your life can be upended by the negligence of distant actors.

Now, let's consider the emotional and societal costs. When data breaches and misuse become rampant, people start to lose trust not just in companies, but in each other and in technology at large. We risk a kind of social withdrawal or inhibition where people share less, connect less, and transact less out of fear.

Already, surveys show that people are withholding information from doctors or financial advisors because they don't want it in a database. That can lead to worse outcomes, such as a doctor not knowing the full picture.

There's also a generational effect. Older folks, who might be less tech savvy, are often targeted in scams using leaked data. They're terrified, and some retreat from using digital services altogether, which can worsen isolation. Younger folks are more comfortable and so share way more, often unaware of how it might bite them later. I occasionally speak at educational institutions, and when I explain how a stupid social post or oversharing personal details can be exploited, the students' eyes widen. They simply had no idea, because the system doesn't teach them to protect themselves in the AI and digital age.

By the end of this chapter, if you're feeling a bit angry or disturbed, I consider that productive. You should be mad that manufactured consent and data profiteering have been normalized. You should be mad that every PII record stored about you is essentially a liability in your name. We've been told a lie that giving up privacy is the price of modern convenience. It doesn't have to be, and in the next few chapters, I'll show how we can turn the tables.

Before we get there, let's touch on one more dimension of the rigging: bias and discrimination. The current systems not only expose everyone to risk; they also disproportionately hurt certain groups. Marginalized communities often get a double whammy.

For example, background checks can contain errors that are more likely to affect people with common names or certain cultural name conventions, leading to false positives such as a mistaken criminal identity. If you have a very common last name, someone else's record might get mixed up with yours. If you're from an immigrant community, your records might not be well integrated and could show gaps or inconsistencies that make you look suspicious.

Landlords and employers, seeing something off in your records, may just pass you by without explanation. This is a rigged aspect, as false or biased data can silently close doors for people. Another angle is that those with fewer resources suffer more from identity theft and from lack of control. They don't have the luxury to take time off to resolve issues, and they can't pay for privacy services or freeze their credit easily and may not even know how. So the inequities of society are mirrored, if not magnified, by the data trust crisis.

Break the Norm

The good news is that we do not have to keep playing by rules that were written without us or for the previous era. Everything you have read so far reveals a pattern created by companies, by regulators, and by us when we clicked "I agree" without realizing the cost. If these pages have stirred something in you, let that feeling become your turning point. The system may have been built to benefit everyone but you, yet that does not mean you must continue through it unchanged. This is your invitation to think differently, choose differently, and begin shaping a future that protects you rather than exploits you.

The system may have been built to benefit everyone but you, yet that does not mean you must continue through it unchanged.

A good place to start is with your own habits. Begin by asking why before sharing sensitive information. If someone wants your SSN, full date of birth, or past addresses, ask what is necessary and how long they plan to keep it. If the answer feels vague or unnecessary, decline.

Choose services that can verify what is required without demanding everything about you. Share only what is needed. Offer a simple confirmation of age, a valid license, or an up-to-date background check instead of handing over documents that expose far more than the situation requires. Separate the parts of your life that matter most from the parts that do not. Use private contact information for finances, healthcare, and employment, and avoid reusing those details for loyalty programs, entertainment apps, or one-time purchases.

Treat your most sensitive information with care, and minimize the number of places it exists by freezing your credit, closing old accounts, removing stored identification, and deleting what you no longer use. The fewer copies of your data that exist, the safer you become. And whenever a company demonstrates respect for your privacy, reward them with your business. Customer behavior drives change faster than regulation ever will.

If you work on a team or lead a business, you have an even greater responsibility and opportunity. Build systems that rely on less data, not more. Replace the default of collecting everything with the discipline of proving only what is essential. Avoid storing raw personal information and instead use tokenized, privacy-first verification methods.

Make privacy choices visible and easy to understand. Ensure that what you collect is genuinely needed and that users have an opportunity to review and correct their information before you act on it. Publish policies that people can actually read, understand, and trust. Support external audits and show your work. Trust grows in the light, not in the fine print.

Responsible enterprises should determine whether their primary mission is to verify identities and backgrounds or to provide their products and services. By regularly asking this crucial question, they will discover effective ways to succeed—avoiding the unnecessary collection of PII for verification or screening, as it falls outside their core purpose—and explore viable alternatives.

Policy leaders and boards can play a critical role by asking one clarifying question before approving any new data practice: *Will this increase or decrease the number of places where sensitive information is stored?* If the answer is increase, pause. Consider a design that reduces risk rather than adds to it. The safest data is the data that never has to be collected at all.

These changes are not theoretical. A landlord can verify that an applicant meets income or background thresholds without receiving an SSN. A marketplace can assure users that providers are verified and safe without storing PDFs in another vulnerable database. A bank can confirm that a license belongs to the right person without keeping a permanent copy. Employers can invite applicants to review their information before decisions are made, saving everyone time, conflict, and frustration. These examples are not futuristic; they are possible today.

> *The safest data is the data that never has to be collected at all.*

The tools exist, and the technology exists. What has been missing is the will to challenge the status quo and the collective will to demand a better approach. The current system counted on our resignation. A new one will be built by our insistence. If enough individuals and businesses choose verification over exposure and control over complacency, the market will change. It has happened before in other industries, and it can happen here.

CHAPTER 2

THE LAYERED COST OF DIMINISHING TRUST

It's tempting to diagnose the crisis of trust as a single issue with a single cause. Maybe it's the dating apps that failed to screen their users, the gig platforms that leaned too hard on convenience, or employers who cut corners during the screening process. But the truth is tougher than any one explanation.

What we're facing is a *layered* breakdown, one that sits at the intersection of human behavior, careless system design, business incentives, and rapidly evolving technology. Solving it requires stepping back and realizing that trust and verification aren't crumbling in one place. They're cracking everywhere at once, and those cracks connect.

Part of the challenge is that each failure looks different on the surface. This is why approaching the trust crisis from a single perspective doesn't work. If you only frame it as a technology flaw, you miss the cultural resignation people feel after years of breaches and scandals. If you only blame corporate negligence, you overlook the reality that even well-intentioned platforms are using tools built for a pre-AI world or merely satisfying regulatory requirements. If you focus only on consumer behavior, you ignore the structural issues that force people to hand over their data just to function.

As I see it, we face a threefold layered challenge that includes structure, societal habits, and technological misalignment. Each layer affects the others, and the erosion of trust is both a technical problem and a social problem. Yes, it's fueled by weak verification standards, but it's also fueled by apathy, convenience, and economic pressure. Think of it like a house with foundational cracks. You can repaint the walls, upgrade the kitchen, install better lighting, and it'll look better for a while. But if the foundation keeps shifting, the problems return.

We face a threefold layered challenge that includes structure, societal habits, and technological misalignment.

That's where we are with digital trust. Companies keep repainting the walls by redesigning apps, adding safety disclaimers, and writing new privacy policies. Meanwhile, the structural flaw remains untouched. We have no reliable, universal way to confirm identity or safeguard personal information before it moves through the system. Until that foundation changes, the surface-level fixes will fall apart.

And there's one more reason we need a multilayered approach: The harm itself is multilayered. When trust fails, the damage spreads outward emotionally, financially, relationally, and sometimes physically. One breach can derail a person's credit for years. One fraudulent hire can jeopardize entire teams. One unverified profile on a dating app can shatter someone's life.

THE FRACTURED LAYERS OF TRUST FAILURES

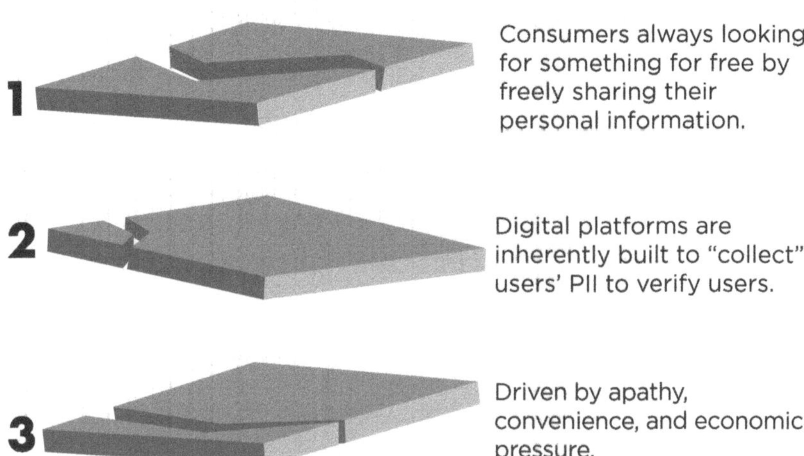

1. **SOCIETAL HABITS**
Years of scandals have caused people to feel that nothing can improve.
2. **TECHNOLOGICAL MISALIGNMENT**
Many platforms are using tools and strategies created before AI was so ubiquitous.
3. **STRUCTURE**
Systems force consumers to hand over their data just to use a platform.

The Cost of Shattered Trust

Imagine a young woman connects with someone on a popular dating app and agrees to meet in person. She thinks the platform has her back and that they've verified this man isn't dangerous. But they haven't, and the date turns into a nightmare when she's assaulted. Investigative journalists later discovered the dating company had known for years that some users were repeat offenders, reported for drugging or assaulting women. Yet, those users easily created new profiles and continued preying on victims.

Now imagine another scenario. A college student signs up for a "free" streaming service trial and casually shares her email and birth date. Months later, that site is hacked. Her personal data ends up for sale on the dark web, where scammers stitch together a synthetic identity, mixing her real details with fake ones, to open fraudulent loans in her name. Before she even graduates, her credit is ruined, and she's fighting debt collectors for loans she never knew about.

These horror stories span completely different parts of life, yet they share a common thread. In each case, there was no reliable way to verify who someone really was before trust was given. The uncomfortable truth is that our society currently runs on a foundation of failed trust. Whether you're meeting a stranger online, hiring a gig worker, or dealing with sensitive information, you're often operating on blind faith. We simply hope people are who they claim to be, because our systems don't give us a better option. There is no reliable way to verify multiple vectors of a human identity without repeatedly disclosing different PII for different purposes.

The uncomfortable truth is that our society currently runs on a foundation of failed trust.

Platforms today either rely on easily faked documents or do no real verification at all. We end up trusting strangers with our safety and personal data by default. The cost of this broken status quo isn't just a few bad actors slipping through; it's a cascading crisis across every marketplace we interact with. From dating apps to gig work, from healthcare to housing, the inability to prove identity and background is causing real people to be defrauded, assaulted, and even killed.

THE MARKETPLACE EPIDEMIC

Think of virtually any online service or marketplace you use, and chances are, trust is fraying there. Let's start with dating apps, which millions of people use to meet partners. Users flock to these apps for connection, but many do so with a knot of worry in their stomach. In fact, 72 percent of online daters say privacy and safety fears deter them from using apps such as Tinder or Bumble.[17]

It's not hard to see why. Dating platforms are rife with catfishing (people hiding behind fake profiles) and stories of harassment or worse. Companies talk a big game about safety, but their follow-through is questionable. One eighteen-month investigation revealed that the parent company of major dating apps had collected "hundreds of troubling incidents" per week, including reports of rape and assault, yet was slow to act on banning or monitoring those users.[18]

Even when apps do ban a predator, it's trivially easy for that person to create a new profile and continue unhindered.[19] In short, the dating industry's trust and safety measures are badly broken, and users know it. They're essentially thinking, *Is this person I matched with even real, and if so, are they safe?* every time they consider meeting someone new. That's an enormous mental burden to carry into what should be a fun social experience.

---

17 "Verified Trust Is the New Currency: Trua's National Consumer Survey Highlights Growing Concerns over Security and Transparency in Digital Platforms," press release, Business Wire, February 11, 2025, https://www.businesswire.com/news/home/20250211045469/en/Verified-Trust-Is-the-New-Currency-Truas-National-Consumer-Survey-Highlights-Growing-Concerns-Over-Security-and-Transparency-in-Digital-Platforms.

18 Lu-Hai Liang, "Platforms Must Prioritize IDV as Trust in Dating, Sharing and Service Apps Wanes," *Biometric Update*, March 27, 2025, https://www.biometricupdate.com/202503/platforms-must-prioritize-idv-as-trust-in-dating-sharing-and-service-apps-wanes.

19 Ibid.

It's not just dating. Gig economy platforms—from rideshare services, such as Uber and Lyft, to home services marketplaces, such as Taskrabbit, Thumbtack, or Care.com—are facing a similar erosion of trust. We are literally inviting strangers into our cars, our homes, our lives. But how do we know those people are trustworthy? The honest answer is, we often don't.

Only 18 percent of Americans feel very confident that popular apps adequately vet the service providers on their platforms.[20] That means the vast majority suspect that background checks and identity verification on these apps range from superficial to nonexistent. They're right to worry because these platforms were never originally in the "trust" business. For example, Care.com is in the business of providing childcare, not identity services. Yet they end up handling huge amounts of sensitive personal data (from SSNs to home addresses) and still can't guarantee that a given babysitter or handyman is who they claim to be.

One of the reasons these platforms are so problematic is because of a faulty, insensitive structure. Digital platforms are fundamentally built to maximize user growth and engagement because their revenue—primarily from advertising or transaction fees—scales with the number of "eyeballs" and interactions. Verifying the real identity, age, credentials, or background of every user can slow their growth, directly threatening the growth metric that investors reward. As a result, the legitimacy of the underlying transaction is treated as secondary, if not incidental, to the business model.

Legally, platforms are shielded by Section 230 in the US and similar intermediary liability protections elsewhere, which explicitly relieve them of responsibility for maintaining trust and safety on their platforms as long as they remain neutral conduits. Platforms,

20 Ibid.

therefore, default to cheap, reactive moderation rather than proactive gatekeeping to weed out unverified or bad actors.

This structural misalignment produces widespread harms: rampant fake accounts and bots, fraud and scams in peer-to-peer services, physical danger in trust-based verticals, and the spread of illicit transactions. The model captures nearly all the upside of network effects while externalizing most of the risks onto users and society. This conflict will persist until liability laws change, revenue incentives are restructured, or verifiable-identity alternatives are available.

Nearly 70 percent of users on home services apps are cautious about sharing personal information with them,[21] and about two-thirds of rideshare users express significant concern about their data security and personal safety on those platforms.[22] We continually hear news stories of rideshare passengers attacked by fake drivers, or gig workers victimized by sketchy clients. People feel that trust, safety, and transparency now outweigh convenience in importance, even if the companies haven't fully caught up. Trust is becoming the new currency.

Online job markets and remote work are another minefield. Millions of jobseekers upload their sensitive résumés and personal details to sites such as LinkedIn, Indeed, or freelance platforms, and 67 percent of users worry about data breaches on these job platforms.[23] But it's not only job seekers at risk; employers are getting duped too.

Scammers have realized that with remote hiring, they might never have to show their face in person, so why not apply with a doctored résumé or even deploy an AI deepfake for the video interview? In one highly publicized case, a man named Soham Parekh allegedly faked

---

21 Business Wire, "Verified Trust Is the New Currency."

22 Ibid.

23 Ibid.

about "90%" of his résumé and simultaneously held multiple remote jobs by gaming the system.[24]

Another fraud ring used deepfake technology to impersonate a candidate and secure a role at a cybersecurity company,[25] a delicious irony, but a costly problem for the duped employer. Background screening today is often a one-and-done, check-the-box process that's easy to cheat and rarely catches ongoing issues. Recruiters on the other end are drowning in AI-generated applications and bots, making it even harder to tell who's genuine.

No wonder only 17 percent of people are very confident that employers have accurate background data on job candidates.[26] Even after you're hired, you might have skeletons in your coworkers' closets that slipped through because no one is continuously verifying anything. It's a trust vacuum on both sides.

Consider housing and rentals as well. Landlords rely on credit reports and background checks that can be riddled with errors. Renters rely on landlords or Airbnb hosts whose identities they take on faith.

That skepticism is well earned. People have been denied apartments or loans due to background report errors they never even got to see, let alone correct. Our current system rarely lets individuals review or dispute their own data in these decisions, leading to awful surprises like "Application denied due to criminal record," only to find

24 TOI Tech Desk, "Here's the Resume of Soham Parekh, the Techie Accused of Working for American Startups at the Same Time," *The Times of India*, last updated July 4, 2025, https://www.timesofindia.indiatimes.com/technology/tech-news/heres-the-resume-of-soham-parekh-the-techie-accused-of-working-for-american-startups-at-the-same-time/articleshow/122225539.cms.

25 Amy Shapiro, "'This Is the Real Long COVID': Remote Jobs, Fake Resumes and Soham Parekh," *CTech*, July 14, 2025, https://www.calcalistech.com/ctechnews/article/byfwjzgilg.

26 "Number of Employers Using Social Media to Screen Candidates at All-Time High, Finds Latest CareerBuilder Study," news release, CareerBuilder, June 15, 2017, https://www.careerbuilder.com/advice/blog/social-media-survey-2017.

out the record belonged to someone else with a similar name. These are life-altering mistakes.

Across every sector, the story is the same. Trust is strained because identity verification is either too primitive or completely absent. And every time trust breaks down, someone gets hurt. People have been scammed out of life savings by impostors.

Patients have received improper care from fake doctors or unlicensed nurses. Families have hired caregivers with dangerous criminal pasts that went undetected. These aren't hypotheticals. They are real cases happening with disturbing regularity. The crisis is so pervasive that 86 percent of consumers factor a platform's reputation for safety and security into their decision to use it.[27]

Across every sector, the story is the same. Trust is strained because identity verification is either too primitive or completely absent.

Trust has become a top deciding factor; if people don't feel safe, they will walk away. In practical terms, that means businesses can no longer assume consumers will just get over a breach or scandal. The public's tolerance for trust failures is approaching zero.

Every Share, Everywhere, All at Once

Why are these trust issues getting worse? One major reason, as mentioned earlier, is the sheer explosion of data breaches in the past decade. We've built a world where everyone's personal information is scattered across hundreds of databases, and those databases are under constant attack. The result has been an epidemic of breaches that

27 Business Wire, "Verified Trust Is the New Currency."

erode trust for both individuals and organizations. We're reaching a point where businesses, in an effort to reduce fraud, continually seek new data signals to verify individuals and transactions. This creates a repeating cycle:

To better detect fraud, companies collect additional personal information (for example, just a few years ago, providing a cell phone number wasn't required, but today it's nearly impossible to sign up or transact on major platforms without phone verification).

These new data points are then incorporated into users' profiles, making those profiles increasingly valuable targets for hackers—and inevitably leading to more breaches, which in turn push businesses to demand even more data to "improve" security.

Consider the scope: In 2023 alone, over eight billion personal records were exposed through data breaches. In 2024, the global cost of these breaches hit $46.25 billion (yes, billion with a *b*) in damages.[28] And that's just the direct, measurable financial fallout. It doesn't include the countless hours of stress and headache for victims trying to freeze credit cards, repair identity theft, or monitor their accounts after each incident.

If you feel like you're getting breach notifications in your inbox every other week, you're not far off. Many of us have essentially become numb to the alerts. In fact, it's become a dark joke that you might have multiple free credit monitoring subscriptions simultaneously because so many companies have spilled your info.

This constant cycle of breach and apology is not normal or sustainable, and people know it. When surveyed, only 33 percent of consumers said they trusted how companies handle and protect their

---

28 *Cost of a Data Breach Report 2025* (IBM, 2025), accessed December 17, 2024. https://www.ibm.com/reports/data-breach.

data.[29] Nearly two-thirds have little faith that when they hand over their SSN or birth date, it will remain safe. Sadly, they're often right to worry, and they still have to hand it over.

It's a vicious cycle. The more we share our data, the more we create attractive targets for cybercriminals, leading to more breaches. Companies today collect more information than they actually need, which only increases the risks unnecessarily and leads to more data breaches than we've ever seen before. Each extra piece of personal data sitting in a database is another lure for hackers and another potential leak waiting to happen.

The more we share our data, the more we create attractive targets for cybercriminals, leading to more breaches.

Deepfakes, Bots, and Digital Deception

As if traditional fraud and human maliciousness weren't enough, we now have to contend with a rapidly rising threat vector: artificial intelligence. Ironically, the same technological wave driving convenience is also supercharging the tools for deception. Generative AI that can create content such as images, video, and text is being weaponized by bad actors in ways that make verifying identity even harder. We've already touched on AI deepfakes being used to fool hiring managers. That's just one example.

AI can churn out thousands of bot accounts or fake applications in the time it takes a human to sip a cup of coffee. It's automating

29 Marc Brodherson et al., "A Customer-Centric Approach to Marketing in a Privacy-First World," McKinsey & Company, May 20, 2021, https://www.mckinsey.com/capabilities/growth-marketing-and-sales/our-insights/a-customer-centric-approach-to-marketing-in-a-privacy-first-world.

fraud at scale. A survey found that 75 percent of consumers fear bots masquerading as real people on online platforms.[30] You might be chatting with what you think is a customer or a seller online, but there's a decent chance you're talking to a software script mimicking human responses. Dating apps are plagued by automated bot profiles "phishing" for victims. E-commerce sites have to distinguish real shoppers from bot armies. Without a better trust mechanism, the AI age risks becoming a fraudster's paradise, where no one can be sure who (or what) is on the other end of a transaction.

Now add deepfakes, AI-generated synthetic video or audio that can make someone appear to say or do something they never did. This technology has advanced so quickly that it's outpacing our ability to detect it. A 2024 report by Regula found that "66 percent of leaders believe deepfakes pose a serious risk to their business."[31] In plainer terms, the very services that banks or employers rely on to check IDs are getting duped by AI-manipulated faces and voices.

There have already been deepfake scams where criminals clone a CEO's voice to authorize a fraudulent bank transfer or impersonate a family member's distressed voice to scam someone. These attacks play on the assumption of trust in what our eyes and ears perceive—an assumption that technology is rapidly undermining.

Alongside deepfakes, the rise of synthetic identities is another AI-driven plague. By piecing together bits of real stolen data with fabricated info, criminals create entirely fake persons that pass credit checks and authentication. Synthetic identity fraud is now one of the fastest-growing forms of financial crime, costing digital marketplaces billions annually. Banks have issued credit cards to "people" who don't

30 Business Wire, "Verified Trust Is the New Currency."

31 Henry Patishman, "The Impact of Deepfake Fraud: Risks, Solutions, and Global Trends," Regula, November 15, 2024, https://regulaforensics.com/blog/impact-of-deepfakes-on-idv-regula-survey/.

exist. Government benefit programs have paid out to millions of phantom applicants. It's happening because our systems still operate on static identifiers (SSNs, account numbers, mother's maiden name) that are either exposed or easily guessed. AI just turbocharges the ability to exploit those weaknesses.

Even when AI isn't being wielded by outright criminals, it's leaking our data in new ways. A phenomenon insiders call the insider threat has emerged, but just not the kind involving rogue employees. Rather, well-meaning employees are inadvertently leaking sensitive data to AI tools.

Picture a lawyer who pastes a draft contract into ChatGPT to "get a quick summary" or a software developer who feeds proprietary code into an AI for debugging help. These folks aren't trying to do harm; they're just using convenient tools. But behind the scenes, many AI platforms quietly collect and retain input data.

That draft NDA or source code could now be sitting on an OpenAI or Google server, outside the company's secure perimeter. Unless you're using a specially configured enterprise version of these AI tools, "inputting personal, financial, legal, or medical information into a consumer-facing AI tool is the digital equivalent of leaving confidential files on a public bench," one article warned.

The employee might have just accidentally handed a trove of sensitive data to a third party without realizing it. And unlike a deliberate breach, this kind of leak is hard to detect because it looks like normal usage. Companies are only beginning to grapple with this, implementing policies such as "Don't paste client data into ChatGPT." But let's be honest, a lot of that cat is already out of the bag. Generative AI has introduced a critical visibility gap for organizations, and data is flowing out to external AI services that aren't designed for secure storage.

All of these AI-driven issues compound the original trust crisis. Fake people, fake content, and inadvertent leaks make the identity verification problem exponentially harder. It's not that AI is *evil*; it's that AI is *amoral*. It will just as readily empower bad actors as good ones, and right now, the bad actors are having a field day. They can attack at scale and speed unlike anything before.

A fraud ring can deploy a thousand bot accounts to phish victims or flood a system, where previously they'd need a call center of scammers. The scale of potential fraud—whether financial theft, misinformation, or reputational damage—is off the charts. We are truly entering an era when seeing is not believing, and any system not built with verification in mind will be gamed. That's the frightening backdrop as we evaluate why we urgently need a new trust paradigm.

Another Layer

There's another layer to this trust crisis that's easy to overlook. While our systems limp along with outdated verification methods, the rest of the world is speeding ahead. Countries, companies, and entire economies are experimenting with new forms of digital identity. Some of these experiments are promising. Others raise tough ethical questions. But all of them point to the same reality. We're heading toward a world where identity will be digital whether we prepare for it or not.

Take Apple's growing push into digital credentials. Millions of people already use their Wallet app to store boarding passes, insurance cards, and event tickets. Several states now allow residents to upload their official driver's license or state ID to the app. Apple has made it clear that the long game is a true digital passport. If that ecosystem continues to expand, it won't just be a convenience feature; it will be a new expectation. The device in your pocket, not a plastic card in your wallet, will become the primary way you prove who you are.

Globally, we're seeing what happens when digital identity is built without guardrails. China's social scoring system shows how identity, data, and state monitoring can merge into something deeply invasive. It's a cautionary tale. When you can link everything a person does to a central profile, you can reward or punish behavior instantly. It's a demonstration of power, not trust. And it's a vision of digital identity that most of the world would never want—but one that still influences the conversation about what verification could and shouldn't become.

Even here at home, debates around voter identification reveal how emotionally charged identity can be. Many fear exclusion, surveillance, or government overreach. But the reality is that over 90 percent of American adults already have a smartphone.[32] If implemented well, mobile identity could actually increase access by removing barriers rather than adding them. Instead of waiting in line at a DMV or carrying paper documents that can be lost or stolen, your verified ID could travel securely in the device you already use for everything else. It is important to note that the objective here is for consumers to have agency over their data with the assistance of the private sector and not any government agency.

Businesses are waking up to the urgency too. Many quietly accept that a certain percentage of their revenue, often around five percent, will simply evaporate every year due to fraud.[33] It's become a line item in the budget. Imagine what would happen if even half of that could be reclaimed. Strong, universal verification wouldn't just

---

32 William Bishop et al., "Mobile Fact Sheet," *Pew Research Center*, November 20, 2025, https://www.pewresearch.org/internet/fact-sheet/mobile/.

33 Michael Cohn, "Organizations Lose 5 Percent of Revenue to Fraud Every Year," *Accounting Today*, April 17, 2020, https://www.accountingtoday.com/news/organizations-lose-5-of-revenue-to-fraud-every-year.

protect consumers. It would meaningfully boost the bottom line for every industry touched by online transactions, which is now almost all industries.

The next wave of commerce will make verification even more critical. Agentic AI systems are already beginning to negotiate purchases, book services, compare prices, and act on behalf of users. This opens an entirely new front of vulnerability. When machines make decisions for us, platforms need to verify not only the identity of the human behind the request but also the authenticity of the agent making it. Otherwise, we will face an avalanche of AI-generated fraud that overwhelms human response times.

The labor market faces similar issues. AI-written résumés, deepfake interviews, and bot-generated applications are flooding hiring pipelines. Recruiters are drowning. Companies are getting duped. Workers with legitimate credentials get lost in a sea of synthetic applicants. The system isn't designed to tell real from fake anymore, and AI is accelerating that gap daily.

In other words, every major frontier—government, business, consumer tech, AI systems, and the job market—is converging on the same pressure point. We need a modern way to verify identity that protects privacy, scales globally, and restores trust before the wave of digital transformation sweeps past us.

"It's Just How Things Are"

With all these threats around us, one might expect consumers to be up in arms, demanding change. And to some extent, they are (as we'll explore in the next chapter). But a large part of the population has responded with resignation.

A kind of learned helplessness has set in. People feel overwhelmed and powerless to fix these systemic issues, so they shrug and say, "Well, I guess this is just the price of living in a digital world." That resignation is perhaps the most insidious cost of all, because it perpetuates the cycle.

We've been conditioned to accept vague privacy policies full of legalese that grant broad permissions, and we do so with a sigh of defeat. Later, when our data is misused or sold behind our backs, the companies point to those very terms and say, "You agreed to this." It's a manufactured consent that we all know, deep down, is not real consent. Yet we keep playing along, because what alternative do we have?

The same resignation colors our approach to data sharing. Especially among younger generations who grew up with the internet, there's a prevalent attitude of *You have to share data to get the service; that's just the deal.* A lot of people (particularly digital natives) shrug off privacy concerns by thinking, *If we want something, we have to provide data for it.*

Older generations, who remember life before everything was tracked, tend to be more alarmed by the loss of control, whereas many younger folks have never experienced true data privacy, so they're less likely to demand it. This generational complacency is dangerous, and society at large has been far too accepting of the status quo.

In our Trua national survey, 78 percent of people agreed that there should be ways to verify identity without constantly exposing sensitive info such as your SSN or home address.[34] This tells us that the vast majority wants a better way. They just don't know how to get from here to there, so they keep handing over their data with eyes half-closed, hoping this time won't be the time it gets abused.

This sense of powerlessness is exactly what the current system feeds on. As long as consumers remain passive, nothing fundamental

34 Trua, *The State of Trust and Safety in Online Marketplaces.*

has to change. It's telling that 91 percent of people say they want more control over how their personal information is used in processes such as loan and job applications and insurance decisions.[35] And 77 percent insist on the right to review their background data before an institution uses it to judge them.[36] These are overwhelming numbers.

People are effectively saying, "We're not OK with the black box anymore, and we want transparency and agency." They see it not just as a preference but as a basic right. In a way, we're facing a digital civil rights issue in our connected age. The latent demand for change is there. The outrage is bubbling under the surface, and you might even say consumers are reaching their breaking point.

Yet, until very recently, there has been a gaping void of alternatives. We want more control, but how? We can't exactly opt out of modern life. Telling someone, "Well, just don't use email, don't use a smartphone, don't use online services" is neither practical nor fair. So people continue forward, uneasily. They share their data because they have to, not because they trust the system. They endure the anxiety of not knowing whom they're really dealing with online, because there's no easy fix at hand.

But it doesn't have to be. This resignation is precisely what we must shake off. We stand at a moment in time where awareness of the problem is higher than ever. Both individuals and business leaders are starting to say, "Enough is enough." And the good news is that solutions do exist.

The first step, though, is galvanizing everyone to admit that the current reality is not acceptable and to demand something better. Stop saying things are good enough. They aren't. We deserve more. The cost of not changing and of having no credential to prove who we are and

35 Ibid.

36 Ibid.

protect our data is becoming too high to bear. Every assault, every fraud, every breach is a stark reminder of that.

The logical question is: *When do we reach a breaking point?* How much more of this can society take before we demand a new way? In the next chapter, we'll explore whether we've already arrived at that decisive tipping point, a moment when the pain becomes so great and the failures so blatant that change is not just possible, but inevitable.

CHAPTER 3

THE BREAKING POINT

To understand how serious this moment is, I always go back to what I consider one of the clearest warning signs we've seen: the rise and fall of 23andMe. For years, this company was held up as a Silicon Valley success story. Millions of people mailed in their saliva to learn about their ancestry, their health, and their genetics. Then everything collapsed.

In late 2023, 23andMe was hit with a massive breach that exposed the most personal data a human being can have.[37] Not passwords. Not credit cards. *DNA*. The genetic profiles of countless users spilled into the hands of people who should never have seen them. When DNA gets exposed, it can be used for things far more dangerous than fraud. It can become a tool for discrimination and surveillance. It can reveal medical predispositions, inherited traits, family connections, and even expose relatives who never even took the test.

---

37 Raj Ananthanpillai, "Bad Genes: 23andMe Leak Highlights a Possible Future of Genetic Discrimination," *The Hill*, October 16, 2023, https://thehill.com/opinion/technology/4254659-bad-genes-23andme-leak-highlights-a-possible-future-of-genetic-discrimination/.

The story got worse when the company went bankrupt. Overnight, that entire genetic database was reclassified as an asset that could be sold to pay creditors. In court, a judge approved the possibility of selling it off. Imagine that. People believed their DNA was sacred and private. But suddenly, it was something that could be auctioned off to whoever wanted to buy it.

23andMe's downfall was a five-alarm fire for anyone paying attention. It forced a brutal question into the open: *When your identity is digitized, stored, and monetized, who actually owns it?* The answer, in this case, was painfully clear. It wasn't the individual. Once people handed over their DNA, they had no control left, and what they thought was a personal wellness tool turned into something that could be weaponized against them.

In 2018, 23andMe sold access to millions of customers' genetic data to GSK for $300 million, giving the pharmaceutical giant exclusive rights to use the information for drug development. Most users were never clearly told that the DNA they submitted for ancestry or health reports could be handed over to a drug company for profit. The consent form simply asked for a broad opt-in to "research," without mentioning that their digital personal DNA would become a third-party corporate asset, and customers received no payment or ongoing control. The overwhelming majority of users agreed to this vague research clause, unaware it could mean their most personal data would quietly fuel a for-profit pipeline.

This lack of clear, up-front disclosure represents a major breach of trust and leaves consumers vulnerable. Once shared, genetic data cannot be taken back—even after 23andMe went bankrupt. Despite promises of de-identification, advances in technology make reidentification possible, raising the risk of future discrimination in insurance, employment, or lending. Millions of people who simply wanted to learn about their heritage unknowingly became unpaid contributors to pharmaceutical research, with no transparency, no compensation, and no real ability to withdraw their DNA from the companies that now possess it.

Their DNA wasn't protected; it was productized. And when the company stumbled, their genetic profiles became just another bankruptcy asset waiting to be passed along to an unknown buyer. Future generations could feel the ripple effects because genetic data touches everyone in a family tree. A stranger could literally purchase insights into a family's medical future.

The public response to the whole saga was fierce because it confirmed what many people had been feeling but couldn't yet articulate. We've been way too casual with our most sensitive information. People are tired of watching their data be mishandled, sold, leaked, and used against them. 23andMe became the breaking point.

The 23andMe situation also highlighted the myth of consent, as the way we currently practice it in the digital age is largely an illusion. Yes, people click "I agree." But no one believes that agreement includes having their DNA sold off in a bankruptcy fire sale. We've been trained to accept the appearance of control even when we have none, and now we're paying the price financially, socially, and in this case, genetically.

The Consumer Revolt

So, are consumers truly fed up? All signs point to yes, as the sentiment has shifted from quiet discomfort to active outrage. Surveys and behavioral data show a clear consumer revolt brewing over data misuse. We've already mentioned the striking survey result that 78 percent of people want verification without constant data exposure, which shows an overwhelming demand for change.[38] But beyond surveys, look at how consumers vote with their feet and wallets.

People are ditching platforms they don't trust. For example, after high-profile privacy scandals in recent years, tech companies such as Facebook saw usage dips among certain demographics, and there's a continuous drumbeat of users saying they are deleting an app over privacy concerns. When major breaches occur, affected companies invariably face lawsuits and customer exodus. It's become common for people to freeze their credit proactively or use privacy-protecting tools such as VPNs and encrypted messaging as a direct response to feeling exploited. Consumers today will absolutely abandon services that can't keep them safe.

One tangible example is that after a string of data leaks in the finance sector, a significant percentage of customers reported switching providers or reducing online activity. Another example is that, following a massive breach at a credit bureau a few years back, there was a surge in people activating credit freezes and fraud alerts, essentially saying, "I trust no one now."

We've reached a point where consumer trust is so low that it's becoming a market differentiator, and companies are starting to advertise how little data they collect or how they use encryption because they know people care. A study by IBM found that over 80 percent of consumers say they'll not do business with a company if they have concerns about its data practices.[39] The age of blind trust is over. People are angry, and rightly so.

But it's not just anger; it's also advocacy. Consumers are demanding new rights and protections. Look at the rise of data privacy laws such

---

38 Trua, *The State of Trust and Safety in Online Marketplaces.*

39 Bhaskar Chakravorti, "Why It's So Hard for Users to Control Their Data," *Harvard Business Review*, January 30, 2020, https://hbr.org/2020/01/why-companies-make-it-so-hard-for-users-to-control-their-data.

as the General Data Protection Regulation (GDPR) in Europe or the California Consumer Privacy Act (CCPA) in California. These didn't come from nowhere and were a response to public pressure. In the US, polls show bipartisan support for stronger privacy regulation. It's one of the few issues that unites people across political lines, because no one likes being exploited or put at risk.

In that sense, consumer demand for change is overwhelming. The message to businesses is clear: Change your practices or lose our business. Online platforms must prioritize stronger security, data transparency, and identity verification and screening, or risk losing consumers.

In a Trua survey, 86 percent of people said a platform's safety reputation is important to their decision to use it,[40] and 58 percent said they're more likely to endorse platforms that have clear safety and security policies.[41] That means if you don't have your trust act together as a company, more than half your customers are actively hesitant to recommend you, and a huge majority is considering safety in their choices.

On the flip side, 60 percent of consumers say they're willing to pay extra for services that offer enhanced security and verification measures.[42] Think about that. People are literally willing to spend money for peace of mind. Trust has become such a precious commodity that it's now a selling point and a differentiator. Trust is the new currency of the digital economy, more than fancy features or even price in many cases.

Trust is the new currency of the digital economy, more than fancy features or even price in many cases.

40 Trua, *The State of Trust and Safety in Online Marketplaces.*

41 Ibid.

42 Ibid.

Another dimension of the consumer revolt is the push for transparency. There's a growing expectation that companies should clearly communicate how they use your data and what they do to protect it. Yet currently, 87 percent of people feel that online platforms fail to clearly explain how personal data is used and safeguarded.[43] Nearly half of older respondents rated platform transparency as "not clear at all."[44] This lack of transparency is interpreted (probably correctly) as companies having something to hide.

People are basically saying, "If you want my trust, show me. Prove to me that you're doing the right things with my data." When that proof isn't forthcoming, consumers assume the worst, which in turn fuels more distrust and more desire for change. It's a feedback loop of dissatisfaction.

In summary, consumers have moved from a state of resigned acceptance to a state of determined demand. They may not know exactly what the solution looks like yet, but they know the current path is unacceptable. They are starting to take their power as customers more seriously, rewarding companies that respect their privacy and abandoning those that don't. They're vocal on social media, quick to share horror stories of scams or breaches, and increasingly supportive of regulations that give them greater rights. This is the tipping point of public opinion. And historically, when consumers en masse demand change, change happens—whether industries like it or not. Businesses would be wise to listen, because the cry for a new trust model isn't going to fade, and it's only getting louder.

---

43 Ibid.

44 Ibid.

Personal Indicators of a Breaking Point

I often say that when it comes to this topic of digital trust, I want to instill guilt in companies that prompts them to step up and do the right thing, and I also want to instill a proper level of fear in consumers like yourself that inspires you to action. Lest you think I'm being overly sensational, let's step back and look at some everyday absurdities that we've all come to accept and why they're glaring evidence that the trust system is broken. You know the system is broken when ...

I want to instill guilt in companies that prompts them to step up and do the right thing, and I also want to instill a proper level of fear in consumers like yourself that inspires you to action.

You have three different credit monitoring services and didn't pay for any of them. They were all freebies offered to you after various companies leaked your data. Company A loses your SSN—they give you one year of monitoring. Company B loses your credit card info—boom, another year of a different monitoring service. After a while, you're juggling multiple logins and dashboards just to keep an eye on your identity, courtesy of corporate America's breaches. Free credit monitoring has basically become the apology currency for negligence.

You can't remember which companies have your SSN at this point. Was it five companies? Ten? More? In reality, if you've had a typical adult life—jobs, bank accounts, utilities, rentals, doctors' appointments—you've probably shared your SSN with dozens of entities just in the past few years. Every time you do, you're rolling the dice on their security practices. The math is brutal: Each time you share sensitive data, you add another risk vector. More shares equals a higher probability that one of them gets breached, and yet, you often have to share your data to get services.

You've been denied a job, loan, or apartment because of an inaccurate background report, and you had no idea there was a problem until it was too late. This one is tragically common. Perhaps a background check falsely claimed you had a criminal record, or a credit report error tanked your score, or an employment verification mixed you up with someone else. Under the current system, these decisions often happen behind a curtain. You should have rights under laws such as the Fair Credit Reporting Act (FCRA) to see and dispute this data prior to a verifying party adjudicating, but in practice, people often learn of the issue only after facing denial or adverse action.

You reflexively click "I agree" on terms and conditions without reading them, because you literally have no realistic choice if you want the service. Every modern consumer has done this. We joke that nobody reads the Terms of Service that is put together by a bunch of lawyers and blessed by regulators. Why don't we read them? Because they're designed to be unreadable—intentionally long, legalistic, and take-it-or-leave-it.

You've begun to suspect that the person you're chatting with on a dating app or social network might not be real. Maybe their photos look a little too perfect, or their responses feel oddly generic. The fact that your mind even goes there highlights how trust has eroded online. People want to meet actual people, not ghosts and phonies. If you've ever hesitated and thought, "Is this profile too good to be true?" you're responding rationally to an environment that makes it easy to lie about identity. This is a system that's broken.

You feel uneasy hiring a babysitter or dog walker through an app because you just aren't sure the platform's "background check" is meaningful. Maybe you do use the app for convenience, but you still find yourself cross-checking the person on Google or social media or interviewing them extensively because you don't fully trust the system. You're not alone. Maybe you heard a news story about a rideshare driver with a criminal record attacking a passenger or a home repair "pro" who was actually a wanted fugitive. The verification processes in place are inconsistent and often inadequate. So you, the user, carry the burden of wariness. It shouldn't be that way.

Any one of the above *You know it's broken when …* scenarios would be bad enough. The fact that most of us can relate to any or all of them is damning. It shows that the current system isn't just failing people at the margins but is failing pretty much everybody, in big and small ways, hundreds of times a year. We've normalized these dysfunctions, but we shouldn't. They are clear signals that we're at the breaking point.

Different Reactions, Same Fate

There's an interesting dynamic when it comes to how different age groups perceive the trust crisis. Older generations (boomers, Gen X) tend to be far more alarmed about what's happening, while younger generations (millennials, Gen Z) often display a concerning complacency.[45] This divide stems from lived experience. Older folks remember a time when social interactions and transactions didn't require handing out personal data left and right.

They've seen the internet evolve from a novelty to an essential utility, and they recall early promises that it would be safe and responsible. Many older people have also been burned by incidents such as identity theft or online scams, so they carry scar tissue. A baby boomer who has had their bank account drained by a phishing attack is going to be understandably paranoid about digital trust going forward.

Surveys bear this out. For example, 91 percent of baby boomers—even higher than younger cohorts—believe online platforms must do thorough background checks on service providers.[46] And when asked about their confidence in platform safety measures, boomers were the most skeptical, with only 39 percent—compared to higher numbers in younger groups—expressing any level of confidence in current safety measures on apps.[47]

For boomers, 85 percent insist on reviewing their background data before it's used, higher than the overall 77 percent average.[48] In short, older people want more control, and they're very aware of the stakes. They've spent a lifetime building credit, reputations, careers, and they are deeply concerned that sloppy data practices or digital fraud could wipe them out in an instant.

---

45 Carl Phillips and Miles Grinyer, "*Trust Across Generations: Different but the Same*," January 11, 2023, https://www.ipsos.com/en/trust/trust-across-generations-different-same.

46 Trua, *The State of Trust and Safety in Online Marketplaces.*

47 Ibid.

48 Ibid.

Now, contrast that with many younger users. A lot of Gen-Zers and millennials have grown up with a fatalistic attitude about privacy and data. They've always clicked "Agree." They've always shared personal moments on social media for the world to see. The idea of not having your info floating around is almost foreign to them.

A segment of younger folks just believes that we have to provide data if we want something out of it. That's the trade-off they've accepted since day one. To them, the internet's benefits have always come bundled with tracking and surveillance capitalism; it's the only reality they've known. This can lead to a sort of dangerous complacency. They might not take basic steps to protect themselves (such as using strong passwords or pausing before sharing PII) because they assume it's futile or not a big deal. We've even seen a narrative of "privacy is dead, you have no privacy, get over it" taking root among younger generations.

However, it would be a mistake to think younger people don't care at all. When asked the right way, they too express desire for change. But they have a higher tolerance for the current state, likely because they haven't experienced the full brunt of the consequences yet. Many Gen-Zers, for instance, haven't tried to get a mortgage or a high-level job yet—moments when background checks and credit reports suddenly become very relevant.

As they age into those life events, they may get a rude awakening. The risk is that by the time they collectively realize how much the system is stacked against them, the problems will be even more deeply entrenched. We have a window of opportunity right now to educate and empower the younger generation to not simply accept the status quo as normal. We should spark a national conversation and challenge everyone to think differently, starting right now.

The good news is that younger people are extremely tech savvy and adapt quickly when motivated. If we introduce better tools, such as user-controlled trust credentials, and explain the benefits, Gen Z could become the biggest champions of a new system. They're comfortable with digital wallets, biometrics, and all kinds of apps. They've also seen the cautionary tales, as many Gen-Zers and millennials watched their parents deal with things such as the Equifax credit bureau breach, or they've seen friends fall victim to Instagram scams. So the seeds of concern are there.

The ideal scenario is to combine the strengths of both groups: the impatience and high standards of older consumers with the open-mindedness and digital fluency of younger consumers.

On the flip side, older generations bring the urgency and maybe even anger that can drive change, but they might struggle with the technology part if it's too complex. So, any solution needs to be accessible and user-friendly enough that a non-tech-savvy senior can use it confidently. The ideal scenario is to combine the strengths of both groups: the impatience and high standards of older consumers with the open-mindedness and digital fluency of younger consumers. Together, that's a powerful coalition that could push the market and regulators to finally overhaul the broken system.

Paying the Price

Thus far, we've focused mostly on consumers—the individuals using services. But what about businesses themselves? One might wonder, if the current system is so terrible, why haven't businesses fixed it already? The answer, increasingly, is that businesses are also bleeding dollars and reputation under the current system. They've been limping along, patching the leaks as they spring up, but the costs are piling high. We are reaching a point where the economics of the status quo simply don't make sense anymore, even for the companies in charge.

Let's talk dollars first. Data breaches cost businesses mind-boggling amounts of money. According to the IBM study referenced earlier, the average breach results in around $4.9 million in direct costs.[49] For big companies, a single breach can run into hundreds of millions. These incidents also knock down stock prices, drive away customers, and invite expensive class action lawsuits.

It's not a question of *if* a given company will face these costs, but *when*. Many organizations are now victims of multiple breaches within a span of years. There are also substantial "hidden" indirect costs, such as customer churn, brand damage, and lost productivity. After a breach, companies have to invest heavily in damage control: paying for PR services and legal counsel, beefing up security, and, of course, giving those free credit monitoring services to millions of people.

It's an incredibly expensive Band-Aid. Businesses also face regulatory fines and penalties. Under laws such as the GDPR in Europe, fines can be up to 4 percent of global revenue for data abuses. Even in the US, regulators such as the Federal Trade Commission (FTC) have started slapping multimillion-dollar penalties on companies for not safeguarding data. So the legal liability side is growing.

Then there's insurance. The cyber insurance market has boomed because companies are recognizing that they can't fully prevent breaches, so they must insure against the fallout. But those insurance premiums are soaring as breach incidents increase. It's gotten to the point where insurers are tightening coverage.

---

49 Ibid.

Companies pay exorbitant liability insurance fees, effectively taxing themselves for their own risky data habits. And guess what? Those premiums go up every year because ransomware payouts are rising. It's a negative spiral: more breaches → higher insurance costs → more overheads for businesses, which ultimately are getting passed to consumers or cutting into profits.

Now, consider the verification paradox businesses face. Because trust is low, they think the answer is to collect more data to verify identities, which in turn increases their risk exposure, leading to more breaches, leading to even more verification measures. For instance, after identity theft increased, banks started asking for things such as out-of-wallet questions. To do that, they tap more data brokers and collect more personal history, which then, you guessed it, can leak or be stolen. It's a snake eating its tail.

Companies collect more data to verify identity, increasing risk, which then increases the demand for more verification. They're stuck in this loop because they're solving the wrong problem. The root cause is too much data spreading around, not too little. But companies in the current model can't see a way out except to double down on collecting data. It's a self-defeating approach that is costing them dearly.

We should also talk about the drag on innovation caused by the current compliance-heavy, breach-prone environment. Companies, especially smaller ones, are terrified of messing up data handling because it could mean lawsuits or regulatory action. So they spend tons on legal reviews, compliance checklists, and cumbersome processes to cover themselves. One executive lamented that businesses end up "spending more on lawyers than on security" in some cases.

And yet, despite all the compliance, or arguably *because* of the false sense of security compliance gives, breaches still happen. So businesses are shelling out money on an ever-growing stack of regulations and audits, but none of those truly solve the fundamental issues of data ownership and accuracy. It's a huge overhead with dubious payoff.

Meanwhile, companies that want to innovate—say, a startup with a cool new functional app—are finding users hesitant to sign up, unless trust can be assured, if it means providing personal info. If trust becomes a barrier to user adoption, that's a business problem. Every breach, every fraud scandal chips away at the overall trust in the digital economy, and trust is the foundation of commerce. If people stop trusting online transactions, businesses lose. It's that simple.

In the worst-case scenario, if we had some massive collapse of digital trust, the economic impact could be trillions of dollars lost. Even short of that apocalypse, we're already seeing a pullback where consumers who won't use certain services, or who demand costly safety measures, are slowing growth. If trust erodes, people may retreat from the digital economy in significant ways. Smart businesses know this and are getting nervous.

Finally, consider the competitive field. We're nearing a scenario where, if one company embraces a safer, privacy-first verification method and another clings to the old ways, the former could gain a huge edge. Consumers will flock to the service where they feel safe. Businesses are starting to realize that first movers on trust will win loyal customers, and laggards will be punished. Businesses are not blind to the shifting winds. Forward-thinking leaders see that the liability is unsustainable and the customer demand for privacy is growing. They know that if they don't adapt, they'll lose out.

When More Rules Don't Help

If businesses are bleeding and consumers are furious, you'd think regulators and governments would have solved this by now, right?

After all, protecting citizens is part of their job. Indeed, over the years, lawmakers have rolled out a slew of regulations aimed at data privacy and security.

But the ugly truth is that our current regulatory approach is a patchwork mess that hasn't solved the core problems. We have what one might call an alphabet soup of U.S. agencies and laws—CFPB, FTC, FCC, CCPA, HIPAA, FCRA, GLBA—each addressing only a slice of the issue, usually after a major failure has occurred. This approach is reactive and piecemeal.

For example, the FCRA was enacted long ago to rein in credit bureaus after abusive practices, the Health Insurance Portability and Accountability Act (HIPAA) was made to protect health data after some high-profile breaches, and the GDPR came after Europeans had had enough of tech companies' data hoarding. Each regulation was well-intentioned, and some have provided important protections. But none of them address the root cause: that individuals aren't in control of their own data.

What's happened is a pile-on effect with new rules on top of old rules and compliance requirements layered ever thicker, yet breaches and identity fraud continue unabated. In some cases, regulation has arguably made things worse by creating complacency. Companies tick the compliance boxes and then consider their duty done—until they get hacked because the regulations tend to be more about process than outcome.

It's largely been security theater. For instance, you might get a privacy policy update email because of some law change, but does that actually make you safer? Not really; it's just paperwork. Meanwhile, fundamental issues such as data minimization aren't really enforced

Regulations tend to be more about process than outcome.

strongly anywhere. Regulators are trying, but these laws are limited in scope and often unknown or unused by consumers. It's telling that despite an ever-thicker rulebook, users still have no real visibility into what data exists about them in various databases, nor an easy way to correct errors preemptively. Usually, you can only react after the harm is done.

From a business perspective, the regulatory landscape is equally frustrating. Regulations are costly to comply with, and the punishments for failure to comply are harsh, but they don't necessarily prevent the failures. It's like being forced to wear a hundred bandages to prevent a wound instead of just curing the underlying disease. Companies end up pouring resources into compliance departments, audits, and legal fees rather than into truly securing systems or innovating safer architectures.

We're trying to force third parties to protect first-party data. In other words, we expect every individual company to perfectly safeguard the piece of your data that they hold, and when they fail, we scold them or fine them. But as we know, this third-party data model, where every company stores duplicate personal data, is inherently vulnerable. It's unreasonable to expect, say, an online elder care platform, a dating app, and a local grocery chain to all have state-of-the-art cybersecurity to protect the slices of PII they each own. Inevitably, some will slip up, and your data is only as safe as the weakest link. Regulators focusing on penalizing the weak links after the fact doesn't stop breaches; it just punishes them postmortem.

We're trying to force third parties to protect first-party data.

When I've discussed a new approach with regulators, they are very keen on the idea. Why? Because they, too, are sick of the never-ending breaches and complaints. Public agencies such as the FTC or the Consumer Financial Protection Bureau (CFPB) are inundated with consumer complaints about identity theft and credit report errors. It's like bailing water out of a sinking boat. They'd much prefer someone patch the hole in the hull.

All this points to a major opportunity. If we introduce a better model that demonstrably reduces breaches and co-opts individuals in guarding their own data, regulators could and should absolutely embrace it without fearing any loss of oversight or a weakened regulatory framework. In fact, it could replace large chunks of the alphabet soup with a simpler, streamlined regime. Instead of trying to enforce privacy by punishing misdeeds, the law could enforce privacy by design, requiring that systems be built in a user-centric, minimal-data way from the ground up. We're not there yet, but we're close to acknowledging that what we have isn't working.

From Reactive to Proactive

While it's tempting to look to government as the answer, it is critical to avoid solutions that invite government control or centralized management of digital trust credentials, as this path has already led to authoritarian social credit systems in countries such as China. While governments remain the primary issuers of foundational identity documents such as passports, national ID cards, and driver's licenses, these credentials were designed for physical world use and are increasingly vulnerable to sophisticated forgery, deepfakes, and synthetic identity fraud.

> *It is critical to avoid solutions that invite government control or centralized management of digital trust credentials.*

As a result, private sector entities engaged in e-commerce, financial services, gig platforms, and online marketplaces have no reliable alternative but to develop and operate their own robust, independent verification and screening layers to confirm the authenticity of these government-issued documents and the legitimacy of the individuals presenting them.

This private sector responsibility is not a choice, but a necessity driven by risk and economics. Enterprises, platforms, and service providers bear the direct financial and reputational liability when fraudulent identities are accepted, whereas governments typically face no comparable accountability for the growing ease of credential forgery. Delegating trust verification exclusively to state actors would not only concentrate excessive power in governments but also fail to keep pace with the speed, scale, and technological sophistication required by global digital commerce.

Instead, competitive, privacy-preserving, and interoperable verification solutions developed by the private sector—using advanced document forensics, biometric binding, and decentralized user control—offer the most effective and least authoritarian path forward.

If the government does want to be more involved, the best thing it can do is alleviate the burden on businesses by changing the structural expectations (more on this later). Right now, the government's approach is that companies must protect the data they collect, and they will be punished if they fail. But if companies didn't collect personal data but used a verified credential system instead, they'd be relieved of certain liabilities.

This would entice businesses not to want the data in the first place, which is exactly what we want! One can envision updates to laws such as the FCRA where, if a decision is made via a user-provided credential, it's out of the old credit bureau scope and thus simpler.

Or consider employment laws. If a hiring decision is made based on a user-owned background check token that the user pre-approved, companies could avoid some of the cumbersome processes and potential disputes. Governments can craft these incentives.

While I strongly believe government should not be the primary driver, it's worth noting that government itself stands to benefit from a trust framework. Think of how many government services rely on verifying a person's identity or background—screening workers for non-sensitive jobs, welfare benefits distribution, and voter verification in elections. Governments spend billions on these processes and still often get it wrong, whereas a Trust Bureau model could streamline a lot of public sector operations.

That's why I often draw the parallel to TSA PreCheck, a government-run program that essentially issues a trust credential to travelers after vetting them, speeding them through security. It was wildly successful in aviation; why not apply similar logic broadly?

In fact, the TSA PreCheck example is a proof of concept that a better system can work. PreCheck showed that people will gladly opt into a vetting process if it means a reusable credential that makes life easier and safer. Millions of Americans paid and enrolled to get that Known Traveler Number, which is basically a trust token for airport security. That's voluntary adoption of a trust credential in a specific context. Now, PreCheck is government-run and specific, but it demonstrates behavior: If the value proposition is clear (less hassle, more security), people sign up.

So what's the holdup? Largely inertia and the structural issue of being stuck in an industrial-age process where a third party is responsible for guarding the consumer's data without any ulterior motives for doing so in the digital age. But the timing is ideal. We're seeing data privacy get attention at the highest levels. While political winds shift, the underlying issue commands attention because it touches economics and national security too. Everyone's worried about Chinese data theft, for instance, which is directly related to our flawed data practices. The government can seize this moment to champion a new infrastructure for digital trust.

Think of it like the interstate highway system or the internet itself. Big innovations often needed an initial government push or standardization. A Trust Credential system might need that kind of boost. But once in place, it would relieve the government of creating a myriad of statutes, regulations, and laws, leading to expanding the government workforce to enforce those laws.

PART II

WHAT IS THE TRUST CREDENTIAL FOR LIFE?

CHAPTER 4

YOUR CREDENTIAL FOR LIFE

The internet began as ARPANET in the late 1960s, funded by the US Defense Advanced Research Projects Agency (DARPA) and designed explicitly to connect a small, closed community of defense contractors, universities, and government researchers. In that early ecosystem, everyone knew one another either personally or by institutional reputation, and machines were physically secured inside university labs or military facilities.

Trust was implicit. You were either on the approved list of nodes, or you simply weren't connected at all. There was no concept of anonymous public access, no commercial traffic, and therefore no perceived need for strong, built-in identity or access-control mechanisms at the protocol level. When the National Science Foundation later opened NSFNET and then the commercial internet exploded in the 1990s, those original design assumptions—high trust, known participants—were carried forward unchanged into a world of billions of strangers.

That absence of native identity has proven to be the internet's enduring Achilles' heel. As every aspect of human life, from banking, health records, and voting to social interaction and critical infrastructure, migrated online, we retrofitted layer after layer of point solutions: passwords, two-factor authentication, CAPTCHAs, reputation systems, verifying biographic details of an individual, solutions to verify biometric data used by some vendors, fraud-detection machine learning, etc.

Each solution addressed a specific threat vector, but none restored the missing foundation of verifiable identity baked into the fabric of the network itself. The problem has only worsened with generative AI, which can now impersonate voices, synthesize biometric data, craft flawless phishing at scale, and automate credential-stuffing attacks faster than defenders can react. What began as an academic trust model has become the single greatest systemic risk in our digital society, and decades of patchwork defenses are finally reaching their breaking point.

A Complicated Patchwork

Today, verifying your identity is a prerequisite for everything from opening bank accounts to accessing online services. Yet traditional verification systems are fragmented, inefficient, and vulnerable. These mechanisms, intended to prevent fraud and ensure compliance, typically operate in silos across different institutions and jurisdictions.

The result is a patchwork of one-off checks that don't communicate with each other or build upon past verifications. A single snapshot verification assumes a static risk profile, ignoring life changes such as a new address or emerging threats such as AI-generated synthetic identities. This static, repetitive approach leaves exploitable gaps, proliferation of personally identifiable information across millions of organizations, and creates compliance nightmares.

US regulations such as the FCRA demand accurate, updatable consumer reports, and the EU's GDPR mandates strict data minimization and consent. Yet, traditional systems often force people to submit the same personal data repeatedly, indirectly contradicting these laws and exposing companies to penalties for noncompliance. In short, the old model of identity verification is past its expiration date. It's cumbersome for users, costly for businesses, and full of cracks that bad actors are slipping through.

The old model of identity verification is past its expiration date. It's cumbersome for users, costly for businesses, and full of cracks that bad actors are slipping through.

If you're a business leader, these flaws translate into real inefficiencies and risks. Repetitive identity checks drain your company's resources and slow down operations. Consider a typical client or employee onboarding process. Each step might require resubmitting sensitive personal information such as SSNs or passport details, scattering that data across multiple databases and multiplying the risk of a breach.

This lack of clarity drives users away, and the duplication also inflates costs. For instance, banks collectively spend billions on manual compliance reviews and audits each year. Errors are common when information is fragmented, leading to false positives or false negatives that can trigger lawsuits or regulatory fines. Cross-border transactions add another layer of complexity, as businesses must navigate different national standards, increasing administrative overheads by as much as 40 percent in some cases because of divergent anti–money laundering (AML) rules and data regulations.

Enough of Antiquated Silos

The problems extend to background checks and identity screening practices, which, in many cases, remain stubbornly outdated. Manual background checks often rely on pulling records from various public databases that are incomplete or out of date. The reports generated can be riddled with errors, confusing two people with similar names or showing old information that is no longer relevant or that has violated state, federal, or local statutes.

Such inaccuracies undermine the reliability of screening results and can unjustly derail opportunities, such as job offers or apartment rentals. They also create compliance headaches, as employers must follow guidelines from bodies such as the Equal Employment Opportunity Commission (EEOC) to ensure background checks don't discriminate, and the FTC can penalize companies for failing to ensure fair and accurate reporting.

But the antiquated, siloed nature of many screening processes makes it hard to meet these standards. In an era of remote work and gig platforms, where verification often occurs without ever meeting the person face-to-face, these legacy methods are increasingly easy for fraudsters to exploit. Criminals armed with fabricated histories or even AI-generated fake credentials are slipping through gaps in the system.

The business impact of outdated screening is significant. Errors in public records or delays in manual processing can stretch hiring timelines from days to weeks. That means higher recruiting costs and lost productivity. A mismatched or missed record could lead to a bad hire that increases turnover or, worse, a security incident, both of which carry heavy costs. Companies also risk lawsuits if their background check process disproportionately disqualifies certain groups or if they report incorrect information that harms an applicant's prospects.

There are real financial penalties too. Regulators have fined firms millions of dollars for reports that turned out to be misleading or incorrect. All of these factors are pushing organizations toward continuous, automated screening solutions as a lifeline. By leveraging real-time data feeds and continuous monitoring, businesses can catch issues as they arise rather than relying on one-and-done checks.

This proactive approach can drastically reduce fraud losses (identity fraud alone cost Americans $43 billion in 2023[50]) by flagging suspicious activity or status changes early. It also lowers liability through instant alerts. For example, notifying an employer if a worker's professional license has expired or if a new criminal record for an employee appears allows the employer to take action.

With the stakes so high, it's clear that business as usual won't suffice. We need a fundamentally new approach to establishing trust online, one that is both faster and more secure, and that puts individuals back in control of their personal information.

A New Solution

With all this said, imagine a solution that combines the speed and convenience of an airport fast pass with the rigorous security of a thorough background check, all while letting you, the individual, decide what information to share. This is the essence of the Trust Credential for Life. A Trust Credential for Life is a single, reusable digital credential that contains proof of your verified identity and background and is safely stored in a secure digital wallet on your phone.

---

50 Christina Ianzito, "Identity Fraud Cost Americans $43 Billion in 2023," *AARP*, April 10, 2024, https://www.aarp.org/money/scams-fraud/identity-fraud-report-2024/.

YOUR CREDENTIAL FOR LIFE

Individual gets to verify their identity and background data for accuracy and obtain their reusable Trust Credential for Life.

"TRUST CREDENTIAL FOR LIFE"

You verify once, and then you can use this credential anywhere, anytime, whether you're starting a new job, signing up for a gig platform, renting a home, or proving your humanity on an online forum rife with bots. Crucially, it puts you in control of your data. You only share the specific details required for a given transaction, nothing more. The Trust Credential for Life concept draws inspiration from the TSA's PreCheck program and the new wave of mobile driver's licenses (mDLs) being rolled out by state DMVs.

A Trust Credential for Life is a single, reusable digital credential that contains proof of your verified identity and background and is safely stored in a secure digital wallet on your phone.

TSA PreCheck is a government-run program that, since 2011, has transformed airport security for millions of travelers. The idea is simple. Travelers undergo a thorough vetting comprising a background check, fingerprinting, and even an interview to prove they're low risk, and in return, they get a Known Traveler Number that lets them use expedited security lanes at airports. No more removing shoes and jackets or unpacking laptops; a quick scan and you're through, often in under five minutes.

Over twenty million people have opted into TSA PreCheck because it saves time and hassle while maintaining safety. It's a great example of "verify once, use many times." TSA PreCheck shows that when you build trust through an up-front investment in verification, you can dramatically streamline repeat interactions. Travelers get convenience without sacrificing security, and the TSA can focus its attention on unknown or higher-risk passengers. Importantly, PreCheck is voluntary, has a fee to get it, and is privacy conscious.

Travelers willingly provide their data in exchange for the benefit, and the program is run by a centralized, secure authority (the TSA) that adheres to strict privacy rules.

The success of PreCheck, with its fast lines and frequent flyer appeal, demonstrates that people are willing to participate in rigorous screening if it leads to a smoother experience down the line. Businesses outside aviation could learn from this model to streamline how they vet customers or employees. In essence, TSA PreCheck proves that trust, once earned, can speed up transactions without compromising safety.

Meanwhile, mDLs are bringing a similarly revolutionary approach to everyday identity verification. An mDL is a secure digital version of your driver's license stored on your smartphone. States such as Arizona, Colorado, Louisiana, and many others have begun issuing or testing mDLs, and as of 2025, more than a dozen US states have rolled out programs, with at least eighteen states projected to be on board by the end of 2025.[51]

These digital IDs use encryption and biometric locks like your phone's facial comparison or fingerprint scan to ensure only you can use them. They let you prove things such as your age or identity contactlessly, by scanning a QR code or tapping your phone, without handing over a physical card. The benefits are big. If you lose your phone or it's stolen, the digital ID can be remotely disabled, which is far better than a lost plastic ID floating around. You can update or renew it instantly online, and importantly, you can choose to share only the information that's necessary.

For example, to buy age-restricted products, an mDL can simply transmit a "Yes, over 21" verification, not your full birth date or

51 "18 US States to Adopt Digital Driver's Licenses by 2025 as Mobile ID Adoption Grows," *Mobile ID World*, August 8, 2025, https://mobileidworld.com/18-us-states-to-adopt-digital-drivers-licenses-by-2025-as-mobile-id-adoption-grows/.

address. Privacy by design is a core feature. Governments see mDLs as a way to improve services and reduce fraud, and users enjoy the convenience of not having to carry a wallet full of cards. The fact that the TSA is now piloting to accept mDLs for airport security screening in certain airports shows how much confidence is being placed in these digital drivers licenses.[52] In short, mobile DLs are turning the traditional model of "show me everything on your license" into "show me only what I need to know."

Imagine What Could Be

The Trust Credential for Life essentially asks, What if we improved and applied the lessons of TSA PreCheck and mDLs to the wider digital world of hiring, online marketplaces, financial services, and beyond? As I see it, a Trust Credential is even stronger than the mDL model.

> *What if we improved and applied the lessons of TSA PreCheck and mDLs to the wider digital world of hiring, online market-places, financial services, and beyond?*

The mDL model cannot do selective disclosure to humans, but only to a machine via NFC/Bluetooth, whereas the Trust Credential for Life lets you share exactly what you want, whether the verifier is a machine or a person. When you tap your phone on an NFC/Bluetooth reader, only the requested details are sent, nothing more.

And if you simply show your phone screen to someone, the app automatically displays just the approved information (for example, a big "Over 21" badge and your photo) while keeping everything else completely hidden. The same credential works seamlessly for both

52 Ibid.

electronic checks and in-person glances, giving you true selective control in every situation. Another major difference between an mDL and the Trust Credential for Life is the ability to add or step up to additional background-related screening, such as criminal history, civil issues, education, professional licenses, sanctions, and watchlists.

The goal is to create a portable, privacy-first credential that individuals can carry throughout their lives, using it whenever they need to prove they are trustworthy—whether that means confirming their identity, background, qualifications, or other attributes—*without* repeatedly exposing all their personal data.

At its heart, this concept reimagines verification as something you do once, thoroughly, and then reuse many times. This flip in perspective gives people back their autonomy. Instead of handing over your sensitive information to every new employer, landlord, or app, which creates countless honeypots of data that could leak, you hold your verified data in your own wallet and grant access temporarily to those who need to confirm something about you. It sharply reduces the unnecessary spread of PII. In an age when data breaches have become epidemic, containing the flow of PII is critical.

This approach also builds a strong foundation of trust that endures. Because the credential is continuously updated and reverified in the background, it never goes stale. Think of it as having a living profile of your trustworthiness that you control. If something important changes—say, you earn a new certification, or, on the flip side, if a previously clean record gets a blemish—the credential will reflect those changes. It's much like how credit bureaus update your credit score. The Trust Credential would update your "trust score" or status. But unlike credit bureaus that operate opaquely, here, you are in the driver's seat.

In the following sections, we'll explore how this combined model works and why it's a game-changer. We'll look at the key advantages demonstrated by TSA PreCheck and mDLs, see how those can extend into broader business use cases, and introduce the idea of a Trust Bureau as the backbone of this new system. Ultimately, this approach promises to cut costs and reduce risks for businesses, help professionals move more efficiently through their careers, and allow consumers to feel safer and more in control in their digital lives.

Imagine a New Future

Under the hood, a lot of innovation makes this possible, and it's important to note that we're not talking about science fiction. Every component needed for the Trust Credential exists and has already been proven in some form. First, there's **tokenization**, a privacy technique that replaces sensitive data with cryptographic tokens. Think of tokenization like writing secrets in code. Your real data, such as names, birth dates, and passport numbers, is stored securely in an encrypted vault, and only random tokens (gibberish strings of code that mean nothing to hackers) are shared or stored externally.

If a hacker breaches a database that uses tokens, the hacker gets nothing but useless scraps. No actual names or numbers to exploit. This is exactly how credit card processors keep your card number and transactions safe.

In the context of the Trust Credential, tokenization ensures that when you present your information, you're usually presenting tokens rather than the raw data itself. For example, instead of giving a new employer your actual SSN on a form, your Trust Credential might send a token that proves your SSN is validated, satisfying the employer's requirement without ever revealing the number. Even if that employer's HR database is later breached, your personal info isn't sitting there in readable form.

Next, the Trust Credential adopts a **Zero Trust** security architecture. Zero Trust is a cybersecurity philosophy that says to never assume that anything or anyone is trustworthy by default. Every access request is verified every time, no exceptions. In practical terms, your Trust Credential is guarded by layers of authentication.

When you use it, the system doesn't just say, "Oh, you unlocked your phone, so you must be you." It will typically require a biometric check, such as facial recognition or a fingerprint scan, to confirm it's really you wielding your credential and not an imposter or a malware program. In my own experience, I've seen too many breaches succeed because someone somewhere trusted the wrong thing, such as a familiar device or a known network, and the attackers slipped in.

Zero Trust means everything is verified rigorously, as if every login attempt is coming from a hostile environment. By designing the Trust Credential with this mindset, we make it far more resilient against unauthorized use. Even if a criminal somehow got hold of your phone, they'd still need to scan your fingerprint or face to use your credentials. And if that fails, the system still wouldn't fully trust the request without further checks. It's like approaching every interaction with healthy skepticism, which turns out to be exactly what's needed to restore trust in a world full of data thieves and AI deepfakes.

Behind the scenes, the Trust Credential is backed by **blockchain technology**, an incorruptible and immutable ledger. Blockchain might sound like hype to some, but here, it's used to ensure that once your identity data is verified and recorded, it cannot be tampered with or altered without leaving a trace. Each piece of your credential is entered into this ledger as a secure, encrypted record, so there is an audit trail for any and all transactions that use your Trust Credential.

Blockchain gives us an immutable trail of truth. Unlike a central database that could be corrupted or hacked, a blockchain-based credential means that trust isn't vested in any one company or government office that could fail; it's spread out, with encryption making it virtually impossible for an attacker to forge your credentials.

This ensures that when someone verifies your Trust Credential, they're checking it against an authoritative source that even the issuers themselves can't retroactively manipulate. It's trust guaranteed by math and transparency, not by faith in an institution.

Finally, there's **biometrics**, specifically liveness detection. This is one of my favorite parts because it tackles one of the newest threats we face: AI-driven identity fraud. In recent years, deepfakes, hyper realistic fake videos or digital personas, have exploded onto the scene. Criminals can now generate a video that looks and sounds almost exactly like you or create entirely synthetic identities by combining stolen data with AI-generated faces.

Traditional identity checks struggle with this, and a static photo or scanned driver's license might not catch that a person isn't real. That's where liveness detection comes in. When you enroll for a Trust Credential, you'd typically do a biometric scan such as taking a selfie video that uses AI to verify you're a live human and that you match the photo on the ID you provided. The system might ask you to blink or turn your head in the video, things a static image can't do, to ensure you're not just a picture being held up.

This process, combined with checks against fundamental government-issued identifying documents such as a driver's license, passport, or national ID, ensures that each Trust Credential is bound to a real, flesh-and-blood person who has been strongly verified. It essentially kills off the threat of someone using a computer-generated avatar or a stolen ID scan to get a credential. In fact, the Trust Credential platform that Trua has developed uses advanced liveness AI that can catch even sophisticated deepfakes. We trained it on countless examples so it can flag the subtle glitches that give away a fake.

As a result, every Trust Credential issued comes with high confidence that you are who you claim to be and that you're actually present when using it. This shuts down a whole avenue of fraud that was starting to scare a lot of people in the security world. With the Trust Credential's biometric safeguards, we can blunt the threat of deepfakes and keep the *trust* in our credentials real.

Bringing all these pieces together—tokenization, Zero Trust security, blockchain immutability, and biometric liveness—yields a new paradigm that is both highly secure and deeply empowering to individuals. Instead of your personal data being scattered across countless company databases, your verified identity and background live in one place under your control.

You don't give away your data anymore. You simply provide proofs or attestations on a need-to-know basis. This flips the power dynamic on its head.

The Technology Behind the Trust Credential

To make this concrete, let's break down, using scenarios from different industries, how the Trust Credential for Life might function in practice.

EMPLOYMENT AND GIG WORK

Today, applying for multiple jobs means undergoing separate background checks for each employer, often duplicating the same steps. With a Trust Credential, you would complete one comprehensive verification process, including identity proofing with, say, a facial recognition match and a liveness test to ensure you're a real person and not a deepfake, plus checks of your education, professional licenses, and criminal record.

Once you have that credential, getting hired at a new company could be as simple as clicking "Share My Credential." The employer receives instant confirmation that you have been vetted, perhaps with a trust score or a green light on required criteria, without seeing your raw personal data.

For example, you could prove you have "no disqualifying criminal history" without divulging any other details of your background. Using cryptographic techniques such as zero-knowledge proofs, the system can answer yes/no questions without exposing the underlying data. As a result, hiring that used to take weeks of waiting for background reports could be completed in hours.

Gig economy platforms could similarly onboard drivers or delivery personnel in a day rather than requiring every worker to undergo a new check for each app. This portability particularly helps groups that often face extra hurdles in screenings, such as freelancers who juggle many short-term gigs or immigrants who lack locally established records. It levels the playing field by giving everyone a carry-anywhere digital badge of trust.

Businesses, for their part, benefit from faster hiring cycles, lower onboarding costs, and reduced legal risk. They no longer need to store piles of sensitive PII since verification can be done by referencing the credential, which means fewer breach liabilities and simpler compliance with laws such as the FCRA or AML rules. In fact, companies can more easily stay within the bounds of regulations because they are asking for credentials rather than collecting data themselves, a model that aligns with privacy by design.

An additional win is that by relying on a shared trust infrastructure, companies collectively harden their defenses against fraud. It's analogous to how TSA PreCheck improved security by concentrating vetting in one robust process instead of using ad hoc checks everywhere. Here, too, verification becomes both easier and more reliable. Even emerging threats such as AI-faked identities are mitigated, because the credential issuance would have caught those through biometric liveness tests and other checks up front, and any suspicious changes would trigger alerts.

BANKING AND FINANCIAL SERVICES

When you open a bank account or apply for a loan, banks must perform extensive Know Your Customer (KYC) and AML (Anti–Money Laundering) checks, which currently means you submit documents and personal data that the bank then verifies and keeps on file. Not only is that inconvenient for you, but it also creates another repository of your data that hackers might target. In addition, it is a point in time verification.

With a Trust Credential, the bank could query your digital wallet to get a token that confirms you've been identity proofed and cleared against relevant watchlists. It's like the bank asking the Trust Bureau, "Can we trust this person's identity and background to meet our compliance requirements?" and getting back a certified yes/no without ever directly seeing, for example, your passport scan or your past addresses.

Because the credential is continuously updated in the background, the bank can also have ongoing assurance that you remain in good standing, something static documents can't provide. This first-party verification model, in which the user consents to share their already-verified credential, could eliminate the need for banks to repeatedly collect and store customer data.

The outcome is that account openings and loan approvals could happen in minutes instead of days, fraud attempts could be slashed by using encrypted, tokenized data that's useless if intercepted, and banks would dramatically reduce their exposure to data breaches. Indeed, fewer data silos means less "surface area" for cybercriminals to attack.

For consumers, it means no more tedious paperwork at every new financial institution and greater confidence that their private information isn't floating around unnecessarily. One major global bank has already piloted a system where API queries to a digital wallet can fulfill KYC checks, resulting in up to a 50 percent reduction in onboarding time and significant savings in compliance costs. The Trust Credential approach would enable that kind of efficiency on a broader, interoperable scale.

HEALTHCARE AND EDUCATION

In telemedicine, verifying a patient's identity and perhaps insurance or medical credentials is crucial for secure and legal remote consultations. Today, that might involve scanning IDs, filling out forms, and sharing lots of personal health information up front. With a Trust Credential, a patient could prove who they are with a quick credential share that might also confirm, for instance, their health insurance coverage without revealing unnecessary medical details.

Doctors would know they're treating the right person and that the necessary prerequisites are met, while sensitive data remains under the patient's control. This also helps comply with privacy laws such as the HIPAA, since minimal data is exchanged and none of it persists on the telehealth provider's side beyond the immediate need.

Similarly, in online education or certification programs, verifying students' identities and qualifications is important to prevent fraud—such as someone else taking a test on a student's behalf or falsified prerequisite credentials. A Trust Credential could store verified education history, certifications, and even proof of exam integrity. An online university could instantly check that a prospective student has a valid bachelor's degree from a certain college by querying the credential rather than requiring transcripts to be mailed and manually reviewed.

The online university could also confirm during remote exam logins that the same verified individual is present each time. This ensures fair access and upholds standards without cumbersome bureaucracy. Because credential updates are continuous, if a professional earns a new certification or their certification expires, those changes quickly propagate to all relying parties. For industries where qualifications are key, this continuous update model closes the gaps that can otherwise persist between an event and when databases get refreshed.

All these scenarios highlight how a Trust Credential for Life flips the paradigm: verify once, use it anywhere your identity needs to be screened. It offers individuals greater freedom from repetition and more control over their information while enabling businesses and organizations to verify faster and more reliably with less data liability.

A DRAMATIC SHIFT

These examples all boil down to a simple but profound shift from *data sharing* to *data proving*. Instead of flooding our digital economy with copies of personal documents and records that live on indefinitely across countless servers, we share a single master credential that proves the documents' authenticity without exposing them.

It is like the difference between showing someone an entire file cabinet of papers versus showing them a single certificate that says everything in that cabinet has been verified. The certificate approach is clearly more efficient and safer. By reducing the spread of personal data, we shrink the attack surface available to bad actors. If each person has their Trust Credential and uses it for various transactions, the companies they interact with no longer need to keep bulky files of personal information.

From a personal empowerment standpoint, the impact cannot be overstated. Instead of being a passive subject of repeated identity and background checks or data sharing, you become an active agent of your own reputation. You curate and present your story accurately and succinctly, rather than being judged by potentially faulty third-party records. Many people have experienced errors in a credit report or background check. Under the old system, you often would not know about it until it harmed you, and you would then be forced to fight to correct it.

With a Trust Credential, because it is first-party and you supply and consent to the data being held, you see exactly what is included, and you participate directly in verifying its accuracy. If something is wrong, you get it fixed through official channels before you share your credential. This reduces errors and bias. Your record reflects *you*, not an outdated third-party database entry that might be mixed up with someone else. It is your life and your credentials, and you finally have a say in how they are used.

Now that we are at the end of this chapter, I hope it is clear what the Trust Credential for Life truly represents. It is a reimagining of digital identity from the ground up, built on privacy, security, and human agency. It turns the current model inside out. Instead of blind trust in countless databases, we have transparent and mathematically assured trust in a single user-held credential. It is your credential for life that is always up to date and travels with you as you go through life in the digital ecosphere.

Once you have it, you should not have to re-prove fundamental facts about yourself repeatedly. You also shouldn't need to sacrifice your privacy for convenience or security. With a Trust Credential, we can have both privacy and security.

I often think about my own children and the world they are inheriting. They conduct so much of their lives online in an environment where trust is fragile and cynicism is widespread. Every week brings news of another breach or scam, another reason to be wary of sharing any data. I do not want them to live in fear that their identities may be stolen or their reputations may be misjudged by an algorithm. I want them to own their digital identities the same way we own our homes or cars—with pride, with control, and with the confidence that they are truly ours.

The Trust Credential for Life is my attempt to build that future. It is a credential, but it is also a commitment: a commitment to technology that serves people rather than the other way around. Trust is deeply human and the glue of society. We lost sight of that in the rush of the digital age, letting trust become mechanized and monetized until it began to unravel. It is time to reclaim it. By giving every individual a private and powerful credential they carry through life, we return to an internet and a world where trust can be mutual and respect for privacy is the norm rather than the exception.

Trust is deeply human and the glue of society.

As we move ahead, the next chapter will zoom out to explore how we can implement this idea at scale. Creating individual credentials is the beginning, but to transform trust on a societal level, we need infrastructure that supports it. Behind every credit card is a banking network. In the same way, behind every Trust Credential is a system that issues and maintains it. We call that system the Trust Bureau, and it is the backbone that would make your lifelong credential possible. With your credential in hand, imagine a network built to ensure that credential is universally accepted and always reliable. That is where we are headed next.

CHAPTER 5

THE TRUST BUREAU

The credit bureau system was built decades ago to answer a crucial question: Can I trust a person to pay me back? And in many ways, it works. It standardized trust in the financial world. If you've ever taken out a loan or opened a credit card, you've benefited from the efficiency of a credit score.

Lenders don't need to personally know you, and they rely on that centralized credit report. But this convenience has come at a tremendous cost to our privacy and control. We pay for it in other ways with hidden data harvesting, errors that can wreck lives, and devastating breaches.

This is the paradox of the credit bureau model. It proved that a centralized verification system can be incredibly efficient and valuable, but it also proved how exploitative and dangerous it can become when individuals have no say. The concept was brilliant, but the execution was exploitative. As a result, we've ended up with a system where your data lives in someone else's vault, and you aren't even allowed in.

Credit bureaus aggregate and control your information, while you are the product being sold. And when they mess up through breaches or mistakes, you suffer the consequences. This centralized model, born in the pre-digital age, is straining under modern pressures.

It's not just the privacy and security issues. It's also that credit data is only one slice of our trustworthiness, and even that slice is handled in an opaque way. As I outlined in earlier chapters, trust in our digital world has been eroding. Now we need to reclaim it.

To do that, we must learn from the credit bureau's successes and failures. We need an alternative approach that keeps the useful core idea—a single trusted source to verify identity/background—but flips the power dynamics and fixes the flaws. That alternative is what I call the Trust Bureau.

What Is a Trust Bureau?

A Trust Bureau, in simple terms, is like a credit bureau for all aspects of your identity and trust, except for one crucial difference: You are in control. A Trust Bureau provides identity verification and background screening as a unified, trusted service, analogous to how credit bureaus provide financial trust data. But unlike those old credit bureaus, the Trust Bureau doesn't turn you into a passive data subject. Instead, it empowers you as an active participant.

A Trust Bureau, in simple terms, is like a credit bureau for all aspects of your identity and trust, except for one crucial difference: You are in control.

Think of how the DMV issues you a driver's license. It verifies your identity and driving qualifications and then hands you a license card. You carry it in your wallet and show it when needed. The DMV keeps a record in case someone needs to verify that the license is valid, but the credential itself, your license, is in your hands. In addition, over the years anyone wanting to quickly verify your identity can use your driver's license as a form of ID, and your specific state-issued DL is accepted as a form of ID nationally.

Similarly, a Trust Bureau would issue you a digital credential called a Trust Credential for Life that you own and carry in a secure digital wallet. The role of the bureau is to validate and vouch for your information, much like the DMV vouches for your ability to drive, but the day-to-day control of your Trust Credential remains with you. This is a hybrid model that combines centralized efficiency with decentralized control.

Your Experience with the Trust Bureau

Let me paint the picture of how a Trust Bureau works in practice. Imagine you decide to get your Trust Credential, a one-time process that will unlock countless doors for you in the future while protecting your personal data and privacy. You begin by onboarding with the Trust Bureau. This might happen through a mobile app on your phone. First, you prove your identity by scanning a government-issued ID, such as a driver's license or passport, and then taking a live selfie video.

Advanced AI in the system checks that your ID is authentic and that the selfie is really you, not a photo or deepfake, but a live human face that matches the ID. This liveness check is critical in an age of AI-generated fraud. The system might ask you to turn your head or say a random phrase so it cannot be fooled by an imposter video.

Once your identity is confirmed, you then consent to various optional background checks of your employment, criminal record, education, professional licenses, and any other information that might be relevant to establishing your trustworthiness for different purposes. All of this happens with your knowledge and agreement. You are essentially saying you want a full background screening on yourself so you can earn a multipurpose reusable and portable Trust Credential.

Now, behind the scenes, the Trust Bureau verification infrastructure begins its work. It connects to authoritative data sources, including public and court records, credit records, education registries, and similar systems, to pull information about you. If this sounds intrusive, remember that it is the same kind of information a prospective employer or landlord would dig up during a background check without you being in the loop.

The difference here is that it is done once, comprehensively, by a trusted intermediary rather than separately by each company you deal with. Unlike a traditional background check company that might operate in a black box, the Trust Bureau is transparent to you. Before any credential is issued, you can see all the data that has been gathered. This step is revolutionary because it puts you, the consumer, in the loop.

If there is an error—for example, if a database pulls up a criminal record that actually belongs to someone with a similar name—you can flag it and have it corrected right away. There are no more unpleasant surprises months or years down the road. In this model, you verify the verifier. Only once you are satisfied that the information is accurate do you proceed.

At the end of this process, the Trust Bureau system generates your Trust Credential and TruaScore, like a FICO score. Your Trust Credential might state that your identity has been verified at a high level of assurance since you provided a real ID and biometric proof, that you have a clean criminal record as of a specific date, that you possess the professional license you claim to have, and similar facts for whatever categories of verification you chose to include. The credential is digitally signed by the bureau and loaded into your digital wallet app, which could sit on your phone much like a boarding pass or your credit card.

THE TRUST BUREAU

Relying Parties & Verifiers

One-Click Trust verification of users via the Trust Bureau

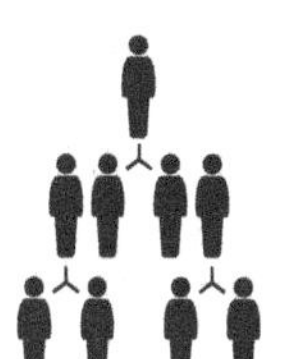

Individuals sign up to obtain their reusable Trust Credential for Life

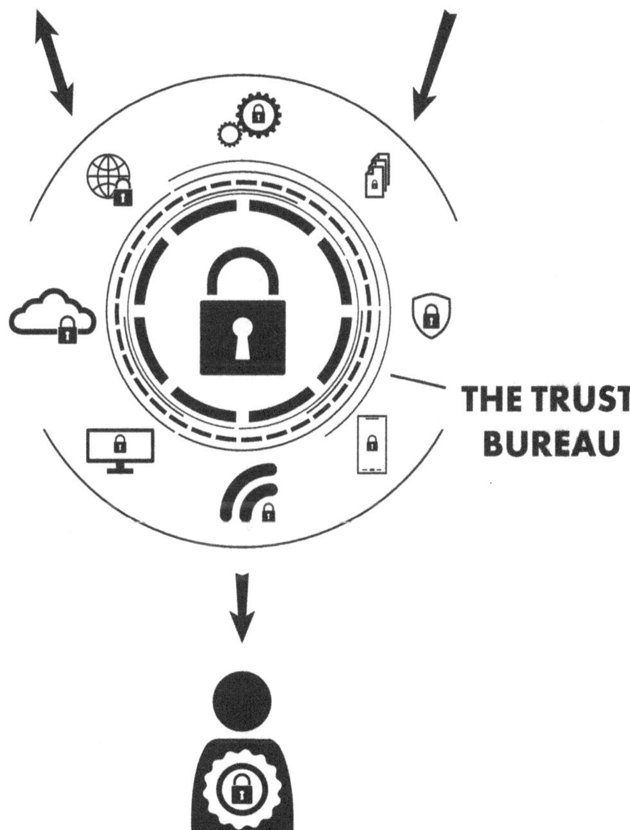

Trust Bureau issues the individuals a Reusable & Portable
"Trust Credential for Life"

Crucially, the actual personal data, your full background report, is not sitting in a central database that every employer can query whenever they wish. Instead, the Trust Bureau gives you more like a token or cryptographic key that proves you are verified and that these are your trust attributes without exposing all the underlying documents each time. The personal data may be encrypted and tokenized on the bureau's backend, but only you hold the keys that allow it to be shared. In fact, the system uses advanced privacy technology, including blockchain and zero-knowledge proofs, to protect your information.

THE ADVANTAGES

With a Trust Bureau in place, those days of repeatedly providing your PII to reverify yourself are gone. Instead, you open your Trust Credential wallet, tap a button to share your credential—or specific portions of it—with the company, and almost instantly, they receive the verification they need. It might be a QR code or a cryptographic token that you send. The employer or platform system then pings the Trust Bureau network with a question, such as whether this person's credentials are valid and whether they meet the stated criteria.

The beauty of this is that the company never sees your raw data; it only receives a verified answer. For example, the query might ask whether a Trust Credential is authentic and whether the user's criminal record is clear over the past seven or ten years. The Trust Bureau system can answer yes or no or provide a specific validated detail without handing over any sensitive personal information.

This uses zero-knowledge proof techniques, a form of cryptography that lets one party prove a statement to another party without revealing the underlying information. You can prove that you are over twenty-one without revealing your exact birth date. You can prove you have a valid driver's license without giving a copy of the license.

This selective disclosure is a major leap forward for privacy. It means companies get exactly what they need to know and nothing more.

From the inquiring company's perspective, the response comes back within seconds and might say that the Trust Credential is valid, issued two months ago, and that this person passes all screening requirements. In that moment, you are approved, onboarded, and ready to go. There is no waiting period and no stack of documents to process.

If you are a gig worker, it means you can start earning money today rather than next week. If you are a renter applying for an apartment, the landlord may see that you possess a gold-standard verified credential and decide to skip running an additional check. If you are on a dating site or home-sharing platform, the site might display a badge indicating that your identity is verified by the Trust Bureau, which helps build trust among other users.

The same single credential can be reused across many scenarios, such as employment, volunteering, online marketplace transactions, travel security, and more. It truly becomes a Trust Credential for Life. Just as your SSN or primary ID is something foundational that you carry with you, this credential becomes something you use whenever you need to prove that you are who you say you are and that you are trustworthy.

UPDATES IN REAL TIME

There is another powerful aspect of the Trust Bureau model: continuous monitoring and updates. Life is not static, and traditional background checks provide a one-time snapshot that captures what is true on the day the report is pulled. What happens if something changes next month?

Perhaps someone earns a new certification, or a college degree, or on the negative side, faces an arrest that creates a record. In the current system, the next company running a check might notice the change or might not, and prior verifications quickly grow stale. The Trust Bureau approach solves this by keeping the credential up to date in near real time. With your consent, the bureau can continuously or periodically recheck key data sources for changes to your status.

It is similar to credit monitoring services that alert you when there is new activity on your credit report, except here, the monitoring focuses on items such as court records or license databases. If a new record appears that affects your TruaScore or your eligibility in some category, the Trust Bureau can update your credential and notify you immediately. This way, relying parties that need to verify your credentials always get the most recent and accurate information, rather than a year-old report.

Just as important, you are never out of the loop. You see those updates, and you know exactly what is being shared. This continuous feature greatly increases trust for everyone involved. An employer can be confident that a lack of alerts means that there has been no significant change because, if something did happen, such as the loss of a required license or a disqualifying incident, the credential would reflect it.

Compare that with today's world, where an employer might have no idea if an employee is convicted of a crime after hiring them unless the employer goes out of their way to run another background check. In critical industries such as transportation or childcare, this difference could literally save lives by preventing dangerous individuals from slipping through the cracks. For individuals, continuous verification can also work in your favor. Positive changes, such as resolving an old issue or earning a new qualification, will boost your trust profile immediately, so you don't have to wait for the next opportunity that triggers a new check.

CAN I TRUST A TRUST BUREAU?

If you're a natural skeptic like me, this brings us to the question that is probably at the forefront of your mind: *Can I trust a Trust Bureau?* This is a fair question, and it's one I will address directly because it goes to the heart of the design. Here are five reasons I believe you can.

First, the Trust Bureau is *not* a rebranded credit bureau. It is a fundamental reimagining of the model for the modern era, and there are specific ways we ensure it does not become a new version of Big Brother. It is innovative and original.

Second, the technical architecture is built for privacy and security from the ground up. The system uses a permissioned blockchain ledger to store data in tokenized form. This means that instead of storing your raw PII in a large central database, it stores encrypted representations, or tokens, of that data on a tamper-resistant immutable ledger.

Only authorized parties with your consent can interpret those tokens. To everyone else, they appear as meaningless symbols. If a hacker were to breach the Trust Bureau systems, they would not find a rich trove of plain personal data to steal. They would find useless tokens and encrypted fragments. In practical terms, your data lives on your device and is in your hands, not in a giant honeypot.

Third, the bureau servers facilitate verification, but they are not hoarding your sensitive files in a way that can be easily exploited. This is a major departure from the status quo, where every company you deal with ends up warehousing your data. We use blockchain not because it is fashionable but because it offers an immutable audit trail and strong protection against tampering.

Every action, including each time someone verifies a credential, can be logged on the ledger in a way that is transparent and cannot be silently altered later. You, the user, can inspect a log of which organizations checked your Trust Credential and when. This brings accountability and transparency to a process that used to happen out of sight.

Fourth, the legal and regulatory structure of a Trust Bureau is designed to avoid the pitfalls of traditional consumer reporting agencies. The FCRA in the United States governs credit bureaus and background check companies. It imposes many requirements on those companies, such as obtaining certified permission and allowing disputes.

The paradigm-shifting approach of first-party (the individual) data sharing often places this model outside the strict scope of that act. The reason is that when you verify your own data, consent, and voluntarily share your verified information directly with a company, the interaction is no longer a company hiring a third party to dig up information about you without your active involvement. It becomes more like you presenting your own credentials, similar to showing your driver's license, which does not trigger the same legal obligations.

This significantly reduces the compliance burden and liability around verifications. Companies do not have to worry as much about the technical procedures of that law or about being sued for background check violations, because the data comes from the individual with explicit consent. Early estimates suggest that this model can cut compliance costs by 30 to 50 percent for employers and platforms by avoiding a large amount of red tape. For regulated data such as health or financial information, keeping it user-controlled can also help with privacy laws.

For example, European regulations guarantee people the right to access and delete their personal data. That can create challenges if you are a company holding thousands of records. If the data is stored in the consumer's own wallet and only tokens sit in the system, compliance becomes easier. The user already has full access and can effectively delete their data by revoking the credential.

Fifth, we are building a governance model and incentive structure that keeps the Trust Bureau honest and accountable. The business model is critical. While traditional credit bureaus make money by selling your data to any lender that pays, a Trust Bureau would make money through user subscriptions and fees for verifications and subsequent authentications. In other words, it functions as a service provider, not a data broker. The data belongs to you, and the Trust Bureau is merely a guardian of that data. It is like a bank vault where you and the bank have a separate key, and you need both to unlock the vault.

While traditional credit bureaus make money by selling your data to any lender that pays, a Trust Bureau would make money through user subscriptions and fees for verifications and subsequent authentications.

That means our incentive is to keep you and the businesses that rely on you satisfied by protecting your data and providing accurate credentials, not to sell more of your data to third parties. If we betray your trust, we lose our users and our revenue. The incentives are aligned. The Trust Bureau succeeds only if it continually earns your trust and the trust of the businesses that rely on it. In addition, I envision external oversight through an advisory council or independent auditors who regularly review bureau practices and publish transparency reports to the public.

The Trust Bureau succeeds only if it continually earns your trust and the trust of the businesses that rely on it.

We could also create a structure in which multiple Trust Bureaus compete under common technical standards. That way, no single entity becomes too powerful, and consumers can choose the provider they trust most. Competition would encourage each bureau to remain ethical and innovative. You could transfer your credential to another provider if one disappoints you, much like switching banks or telecom providers.

I recognize that these governance ideas describe the future, but they are important safeguards that prevent the concentration of power we saw with credit bureaus. The core idea is that you are no longer the product. You are the empowered customer. The Trust Bureau works for you, not the other way around.

COMPARE THE ALTERNATIVE

While no system will ever be perfect, compare a Trust Bureau to the status quo we have today. In a traditional credit bureau or background check scenario, data ownership lies with the bureau or verifier. They hold large databases with your information. In the Trust Bureau model, you own your data. It is stored in your personal wallet and shared only when you choose.

In the old model, companies often collect information about you without your knowledge. They do it in the background, and your consent is either assumed or hidden in fine print. In the Trust Bureau model, nothing happens without your clear and explicit consent for each sharing instance. Consider visibility. With traditional credit reports, you usually see the information only after it has been used, such as when you learn about a credit pull or an error after being denied a loan.

With a Trust Bureau, you see everything first. You pre-approve the data, and you can view a dashboard that shows each verification. Traditional systems are reactive and slow when fixing errors or updating information. You have to file a dispute and wait, whereas the Trust Bureau is proactive and aims to catch errors during initial verification, allowing you to correct data in real time before sharing it with relying parties or verifiers, preventing potential harm.

If something does slip through or changes later, you are notified and can address it quickly. In short, this flips the dynamic from a paternalistic idea that says "trust us, we know what is best for your data" to a collaborative form of trust that involves you at every step. The Trust Bureau enables verification with your participation, making the entire system far more trustworthy.

Now consider the practical benefits to businesses and society once such a system is in place. For businesses, a Trust Bureau can be a tremendous advantage. Today, every company feels compelled to collect far more personal data than it truly wants, simply because it needs to verify identities or perform background checks to provide goods and services or employment (full time or a gig). A caregiving platform does not really want to handle passports or SSNs. That is not its core mission. Yet it has to collect those documents to ensure trust and safety on its digital platforms.

This turns the company into an accidental data hoarder, with all the associated risks. If that platform could rely on a Trust Bureau, it could verify credentials without ever touching raw personal information, eliminating those dangerous data stores. That means a dramatic reduction in breach risk. It's simple. What you do not have cannot be stolen. Companies that adopt a Trust Bureau model could reduce the sensitive data they store by as much as 70 to 80 percent, based on our analyses.

Fewer stored records means fewer targets for hackers, which in turn lowers the likelihood of costly breaches. Imagine being able to tell your customers that you are not holding their birth date, SSN, or other highly sensitive details and that you verify what you need through a secure third party and never retain it. That is a compelling message. It also affects the bottom line directly because lower risk can translate into lower spend on technology to store and guard any data as well as lower cybersecurity and other liability insurance premiums. Insurers calculate premiums based on exposure and potential loss.

If a company can demonstrate that it does not store large volumes of personal information and that verifications run through an ultra-secure tokenized system, insurers will reward that profile. Adopting a Trust Bureau could cut cyber insurance premiums by an estimated 20 to 30 percent for mid-sized firms, saving tens or hundreds of thousands of dollars each year. There is little reason to pay to insure a risk that you have largely eliminated.

Another major benefit is the speed and efficiency of onboarding. Many people have experienced or heard about background checks that take weeks. A job offer might be contingent on a report that keeps getting delayed, or a gig worker might have to wait to start driving because a screening vendor is slow. These delays cost businesses money. They also frustrate strong candidates who may withdraw and go elsewhere.

Today, nearly all employment background checks in the US are performed by third-party consumer reporting agencies (CRAs) rather than by employers directly. Because these agencies compile and sell personal data about individuals, they are classified as consumer reports under the FCRA and are heavily regulated by the FTC and CFPB. Employers must follow a strict multistep process by providing clear written disclosure, obtaining explicit written authorization from the candidate, waiting for the report, sending a pre-adverse action notice if something negative is found, waiting again, and only then can they withdraw an offer. On top of that, the FCRA explicitly prohibits running a background check before a conditional job offer is extended in most cases, adding weeks of delay and friction to hiring.

With Trua's Trust Credential for Life, the entire model flips to first-party, consent-driven verification. The individual owns and controls their own verified background data inside their digital wallet. At any point in the hiring process, the candidate can choose to share exactly the relevant portions (for example, "no felony convictions in the past seven years") directly with the employer. Because the data is shared peer to peer with explicit, on-the-spot consent and is never assembled or sold by a third-party CRA, the transaction falls mostly outside FCRA regulation. Employers get instant, accurate answers; candidates maintain full control and privacy; and hiring can move as fast as both parties want.

By using reusable Trust Credentials, companies can shorten onboarding times from days or weeks to minutes. This is not an exaggeration. When someone arrives with a verified credential, the verification step becomes a quick formality. For roles with high turnover or urgent staffing needs, this speed is crucial. Think of a rideshare company onboarding drivers. Instead of telling an applicant to start next week after the background check clears, the company can say to start that afternoon, because they have already verified the applicant's Trust Credential.

That kind of agility can be a competitive edge. In the gig economy alone, which includes tens of millions of workers, the savings in time and money are enormous. A driver who has been verified once can carry that trust to every platform, whether it's ridesharing, food delivery, or task services, without each company having to repeat the same work. This not only cuts costs but also increases workforce mobility and access. A small startup can onboard a trustworthy freelancer just as quickly as a large company, which levels the playing field.

From a compliance standpoint, companies also gain peace of mind. Using a first-party, user-consented credential means fewer situations where a company handles data in ways that trigger heavy regulatory scrutiny. For example, under United States law governing credit reporting, if an employer uses a third-party background check, it must follow strict processes around notifications and adverse action letters when deciding not to hire someone based on a report.

When the information comes directly from the candidate through a Trust Credential, that process can be simpler, since the candidate has effectively pre-vetted and shared their own information. The Trust Bureau also standardizes the process with built-in compliance. It ensures that consents are in place, records every check to provide an audit trail, and keeps data current. All of this can reduce legal disputes over background checks. Today, companies are frequently sued for technical violations or data breaches. A Trust Bureau can significantly reduce litigation risks by taking over the heavy lifting in a consistent, lawful manner.

If an individual's data is incorrect, that person can correct it up front, preventing a company from unknowingly making a decision based on faulty information and then facing a lawsuit later. The entire system is more defensible legally because it is transparent and permission based.

Imagine What Could Be

Now, step back and imagine the wider societal impact if Trust Bureaus and Trust Credentials for Life became common. We would see a dramatic reduction in identity theft and fraud because there would be far fewer weak points. In recent years, Americans have lost many billions of dollars to identity theft and related fraud.

A major cause is that personal data is scattered across countless locations, waiting to be stolen. By sharply reducing the spread of PII, a Trust Bureau helps plug those leaks at the source. Even when breaches occur, the damage is limited because stolen tokenized data cannot easily be used to impersonate someone.

A portable Trust Credential could also expand access to opportunities. With a Trust Credential, a user can prove their identity and background in alternative ways. For instance, showing a clean criminal and employment record may reassure a landlord even when the applicant has no credit history. A young gig worker with a short résumé can still present a solid verified profile to a new platform.

By standardizing trust verification, we level the playing field and reduce bias. This helps those who often fall through the cracks of today's system, such as gig workers, immigrants, or young people just starting out. Everyone has the chance to present an authentic, verified self rather than being judged purely by credit scores or scattered data points. It is similar to how the introduction of modern credit scoring opened up lending in the twentieth century. Suddenly, a young person with a short history but responsible habits could access credit, whereas previously they might have been shut out. The Trust Bureau extends that concept to all aspects of trust, not just creditworthiness.

I also believe that this model will push the industry toward consolidation and higher standards. In the past, credit bureaus displaced many smaller and less reliable local agencies by providing unified national systems. In a similar way, the Trust Bureau can replace the chaotic landscape of dozens of identity verification startups and background check firms all following different methods. Many of those providers rely on outdated techniques or have inconsistent compliance.

If a true Trust Bureau takes hold, those players will need to integrate or risk becoming obsolete. That may be difficult for them, but it benefits society. We do not need twenty different companies each storing a copy of your driver's license and repeating the same checks. We need one trusted pipeline that everyone can rely on.

This consolidation around a secure, privacy-first standard will save significant resources. Enterprises could collectively save billions of dollars by eliminating now-redundant compliance work and data breach cleanup costs. Those savings can translate into lower service costs or higher wages, since gig platforms will incur less fraud-related loss. It leads to a more efficient ecosystem overall.

Rebuilt Confidence

On an emotional level, and perhaps most important of all benefits, a Trust Bureau can help rebuild public confidence in the digital world and institutions. Much of the anxiety people feel, the mistrust in how data is handled, could be eased if there were a visible system that clearly puts individuals back in control.

I often say that this work is about reclaiming trust, and I mean that quite literally. The Trust Bureau provides individuals with the tools to reclaim control over their personal information. It treats trust as something you curate and carry with you through life, not something produced about you in secret by others. I find that vision deeply empowering. It also provides timely responses to new threats arising from AI-driven fraud. Deepfakes and synthetic identities exploit weaknesses in current systems. They rely on the fact that digital identity checks can be fooled and that data is widely available to steal and misuse.

By incorporating biometric verification, liveness tests, and continuous monitoring, the Trust Bureau makes it far more difficult for a fake persona to slip through. It is a dynamic defense. As deepfakes grow more sophisticated, the Trust Bureau can update its detection algorithms and keep the trust network protected. We are bringing together the best of modern technology, such as AI, cryptography, and blockchain, and aligning it with a philosophy that centers on human beings.

Throughout my career, from helping implement TSA PreCheck to working on national security vetting systems, I have seen how much good can come from finding the right balance between security and convenience. The Trua Trust Bureau, or TTB, as we call our implementation, is the culmination of those lessons.

It combines the proven idea of a credit bureau, a one-stop hub for trust data, with the spirit of individual empowerment that our era demands. It is the foundational infrastructure needed to make the Trust Credential for Life both possible and widely used. Without a Trust Bureau, a personal trust credential would likely remain a niche concept, useful for a single company or limited purpose.

With a Trust Bureau, that credential becomes universally portable and widely accepted because everyone trusts the source. Unlike the creation of credit bureaus in the early twentieth century, which took place without much public input or awareness, we now have the opportunity to build this new system intentionally from the beginning so that it is private, secure, and fair.

As we move into the next part of this book, we will explore what it will take to implement and scale this vision, how businesses can rally around it, how it might be governed, and how each of us can participate in reclaiming our trust. I want to close this chapter by emphasizing both the stakes and the promise.

PART III

HOW EVERYTHING CAN CHANGE FOR THE BETTER

CHAPTER 6

THE BUSINESS REVOLUTION

Because of the Trust Credential for Life and Trust Bureau, I'm convinced we stand on the verge of a tremendous breakthrough in business revolution. This is desperately needed and long overdue.

The cruel irony is that many companies get breached while trying to protect and verify their customers. They thought they were storing all that personal data to serve their users and keep them secure, but in doing so, they created a treasure trove for attackers. The very act of companies handling and holding sensitive data became their Achilles' heel.

If you're a business owner, I want this book to be a dramatic wake-up call. You *should* feel a measure of personal responsibility for the good of your company, the good of the consumers you serve, and the good of society in general. And if you are currently putting your users' data at risk, I would argue you *should* feel a sense of guilt.

That said, while we often frame privacy and security as ethical duties, which they are, they are also bottom-line issues. Breaches cost money. A global survey by Gemalto noted that 70 percent of consumers would stop doing business with a company after a data breach.[53] In other words, a single breach can literally be an extinction-level event for a business, vaporizing customer trust and loyalty overnight. Slow, clunky verification processes cost money. Compliance failures cost money. Lost customers and reputational damage cost a lot of money.

A single breach can literally be an extinction-level event for a business.

Conversely, protecting your customers' data, streamlining their onboarding, and earning their confidence can make you money through loyalty and efficiency. Companies that embrace the Trust Credential and Trust Bureau will save costs, reduce risks, and accelerate growth, while those that don't will increasingly bleed expenses, face regulatory wrath, and lose customers to more trustworthy alternatives. This is a high-impact competitive differentiator, as those who adopt earliest will gain the greatest strategic benefit.

For businesses, the benefits fall into a few key areas: eliminating liability, gaining efficiency, building customer trust and loyalty, attracting new customers, and automating compliance. Let's take a look at each in order.

Benefit 1: Liability Elimination

Every record of PII that a company stores is like a live grenade rolling around your organization. You hope it never goes off, but if it does, the explosion in the form of a breach can be devastating and lead to bankruptcy for many.

---

53 "Majority of Consumers Would Stop Doing Business with Companies Following a Data Breach, Finds Gemalto," press release, Thales, November 28, 2017, https://www.thalesgroup.com/en/news-centre/press-releases/majority-consumers-would-stop-doing-business-companies-following-data.

Right now, cyber insurance premiums are soaring because insurers see how likely and costly data breaches have become. From the insurers' vantage point, it's not *if* you'll be breached, but *when*. The average cost of a data breach in 2023 hit an all-time high of $4.45 million, according to IBM's annual report,[54] and the global aggregate damages reached into the tens of billions of dollars. So imagine the relief you could bring by telling your CFO you can remove most of that risk entirely.

By using the Trust Credential model, companies can drastically shrink the amount of sensitive data they hold. If they're not storing customers' driver's licenses, SSNs, or background check reports because they simply query the Trust Bureau when needed to screen and/or authenticate individuals, then even if their systems are attacked, there's very little of value to steal.

Hackers may not even bother targeting a business when they know it holds no juicy data, so that business suddenly goes from an attractive vault to an empty safe. When credentials are tokenized and no raw PII is stored, essentially, that means no breach liability. It shifts the mindset from "we have to guard this mountain of data 24/7" to "we deliberately *don't* keep data, so we don't need a huge budget for technology and data guardrails, a compliance department, or cyber and liability insurance because we're not a worthwhile target."

It's a bit like carrying all your cash on you, making you ripe for mugging, versus carrying none and using digital payment when needed. If someone tries to rob you and you literally have no cash, you're not going to lose anything. Similarly, a company that leverages

---

54 "IBM Report: Half of Breached Organizations Unwilling to Increase Security Spend Despite Soaring Breach Costs," *IBM*, July 24, 2023, https://newsroom.ibm.com/2023-07-24-IBM-Report-Half-of-Breached-Organizations-Unwilling-to-Increase-Security-Spend-Despite-Soaring-Breach-Costs.

trust tokens instead of raw data doesn't need to fear hackers because there is nothing of significance for them to hack.

Reducing breach incidents not only avoids direct costs but also spares you the massive hit to your reputation and customer trust, which is even harder to win back. It's hard to put a precise monetary figure on brand damage, but we know it's huge. By eliminating the root cause—data hoarding—you essentially remove the explosive from the grenade.

Benefit 2: Efficiency Gains

The Trust Credential can fundamentally streamline some of the slowest, most cost-intensive business processes—notably, hiring and customer onboarding. Let's talk hiring first, because nearly every business feels this pain.

According to a recent IBISWorld report, over a hundred million background checks were performed in the United States in 2024, and this number is expected to increase rapidly.[55] The current state of hiring is often an applicant tsunami, especially in the era of easy online applications and AI-generated résumés. Post a job opening, and you might get hundreds or thousands of applications, many from unqualified or fake candidates. HR teams drown in this influx, spending weeks sifting through résumés, conducting background checks, and waiting for third-party reports to come back. A quality candidate might be stuck in an eight-to-twelve-week pipeline from application to hire, during which time they could get an offer elsewhere and disappear. That's a lost opportunity for the company.

55 Connor Zaminski, "Background Check Services in the US - Market Research Report (2015-2030)," IBISWorld, last updated September 2025, https://www.ibisworld.com/united-states/industry/background-check-services/6058/.

Now consider the Trust Credential model. Candidates apply with their Trua-certified digital Trust Credential attached. Instantly, your system can filter out bots and fake applicants because only verified humans with authentic credentials are coming through. You're not left guessing if a résumé is submitted by AI or if the applicant is a real human being because the resume and the application are accompanied by a credential that has already done the vetting.

That alone can cut the pile down dramatically. Then, you can prioritize candidates by trust score or by specific verified qualifications. Instead of spending the first few weeks verifying and screening, you jump straight to interviewing the most promising real candidates that meet your job requirements.. And guess what? There's no waiting for a background check at the end, because it's already in their credential. You essentially hire them with a verified background in hand.

Companies that piloted this approach saw time-to-hire drop to one to two weeks in some cases.[56] Roles that used to take two or three months to fill were getting filled in a matter of days. The savings from that speed are enormous. Every day a position goes unfilled is productivity and revenue lost. One study estimated that a vacancy can cost companies around $500 a day in lost output for a typical mid-level role.[57] Fill it six weeks sooner, and that's $21,000 saved on one hire, not to mention the opportunity cost of whatever that employee will produce in those six weeks.

In the initial phase, while the Trust Bureau is being populated with screened and vetted candidates, verification costs will continue to be borne by sponsors or employers—consistent with current market

56 This is based on some of our customers implementing Trua (www.TruaMe.com) solutions.

57 "The True Price Tag: Understanding the Costs of Bad Hires and Open Seats," Partnership Employment, accessed November 26, 2025, https://partnershipemployment.com/the-true-price-tag-understanding-the-costs-of-bad-hires-and-open-seats/.

practice. Once a critical mass of individuals has been onboarded, these costs decrease dramatically, as subsequent verifications are low cost to verifiers/relying parties and near zero for individuals.

With Trust Credentials, the long-term cost structure fundamentally improves. Individuals maintain their own portable, verified credentials (typically at little or no direct expense, as they are often subsidized by ecosystem participants). Employers or other relying parties will pay a modest query fee to access the Trust Bureau, but this fee is negligible compared to today's time-intensive repeated human identity verification or background checks, document chasing, and manual report interpretation. The result is near-instant, unambiguous verification that eliminates administrative burden and allows HR and hiring teams to focus on higher-value activities.

Now extend that efficiency to customer onboarding. Banks, for example, constantly face the friction-versus-fraud trade-off. Tighten onboarding to prevent fraud, and you drive good customers away with cumbersome paperwork and verification hoops; loosen it, and you invite fraud or compliance issues. Trust Credentials let you achieve both security and a smooth user experience.

A bank using this approach could onboard a new customer in minutes. The user shares their Trust Credential, which already has KYC information verified, perhaps even their credit profile or references. There's no need for the customer to manually fill out forms and upload documents—their credential is shared and it's done.

This speed not only delights customers, but it also means revenue starts flowing sooner. In industries such as financial services or telecom, getting a customer through the door even a week sooner can mean earlier revenue capture and less drop-off during onboarding.

We've touched on this before, but consider gig platforms and marketplaces. Currently, there are over sixty million people in the gig economy, and this number is growing.[58] This sector has high turnover and frequent job changes. Right now, if you sign up to be a rideshare driver, a home-share host, or a tasker for a gig marketplace, you often wait days or weeks for approval while the background checks and verification process are completed by the gig platforms. That's time not generating revenue for gig workers or the platform. Alternatively, many platforms skimp on vetting individuals due to current cumbersome verification processes resulting in bad actors' exploitation.

A fast-trust system could reduce that lag to near zero. You can sign up today and start earning right away since your Trust Credential already says you're good to go. That makes the platform more attractive to workers and enables the company to rapidly scale caregivers and service providers when needed, without compromising safety.

It's easy to imagine a race to the top among platforms. Those who advertise that applicants can skip the wait and start earning immediately with their Trust Credential would attract more workers than those stuck in the old, slow model. Similarly, for customers, an online service that says, "We never store your personal data, and your info stays with you via a Trust Credential," will attract privacy-conscious service providers and undercut competitors that demand a lot of data up front.

Benefit 3: Customer Trust and Loyalty

We've already seen how breaches drive customers away. Conversely, being trustworthy can become a market differentiator that wins customers. There's plenty of evidence that consumers are hungry for

58 Ben Winck, "Gig Work Value Is Too Great to Rush a US Overhaul," *Reuters*, last updated May 12, 2023, https://www.reuters.com/breakingviews/gig-work-value-is-too-great-rush-us-overhaul-2023-05-11/.

safer, more transparent digital experiences. In a PwC survey, about 85 percent of consumers said they wish companies would give them more control over their personal data.[59]

Being trustworthy can become a market differentiator that wins customers.

People are sick of data scandals and feeling exploited. So imagine being able to tell your customers you value their privacy so much that you've designed your systems to never hold their personal data unnecessarily. That you use a Trust Credential system, which means you don't keep their information on your servers. It stays under their control, and you only see what you absolutely need, with their permission, to serve you. That's a powerful message. It flips the script from companies being data hoarders to companies being data guardians on the user's terms.

I can even imagine full marketing campaigns built around this idea. A bank could run ads telling customers that their data stays in their hands, not in the bank's databases. An e-commerce company could reassure shoppers that they can buy with confidence because even if the site gets hacked, their personal information can't be stolen since the company doesn't store it. It sounds almost utopian, but the Trust Credential model makes it completely achievable. And the timing is perfect, because every new high-profile breach erodes public confidence even further.

People are already moving toward brands that prove they respect user data. That's not a guess, as survey after survey shows that consumers spend more and stay more loyal when they trust a company's privacy practices. One study from Ping Identity found that 81 percent of

59 Toby Spry and Phil Regnault, "Delivering the Privacy Experience Customers Seek," PwC, February 10, 2021, https://www.pwc.com.au/digitalpulse/customer-experience-trust-privacy.html.

consumers would stop engaging with a brand online after a breach. But if your company never has a breach because it never holds sensitive data in the first place, that same 81 percent becomes a built-in advantage.[60] You've essentially insulated your customer relationships from the trust-shattering events that take down your competitors.

This also changes how companies handle reputation damage when something goes wrong. If a breach does occur, an organization using the Trust Bureau model could truthfully tell the public that attackers may have gotten in, but all they accessed were useless tokens, not anyone's real personal data. Compare that to the typical headline screaming that a company leaked ten million SSNs. One version becomes a temporary embarrassment; the other becomes a career-ending disaster for executives.

Regulators respond differently too. They're far more lenient when a company can demonstrate that it used advanced protections and that no actual consumer harm resulted because the stolen data was encrypted or tokenized. In many jurisdictions, a breach notification isn't even required if the compromised information was properly encrypted because the law doesn't consider encrypted data an actual exposure. Tokenization through the Trust Bureau functions as encryption at scale, sparing a company the enormous costs of breach notifications, fines, and the long tail of reputational fallout.

---

60 "81% of Consumers Would Stop Engaging with a Brand Online After a Data Breach," press release, Ping Identity, October 22, 2019, https://press.pingidentity.com/2019-10-22-81-of-Consumers-Would-Stop-Engaging-with-a-Brand-Online-After-a-Data-Breach,-Reports-Ping-Identity#:~:text=,online%20following%20a%20data%20breach.

Benefit 4: Talent and Workforce

Beyond hiring efficiency, there's also a broader business advantage to being a leader in trust, because you naturally become a talent magnet. In today's market, especially in tech and knowledge-driven fields, people care deeply about a company's values and reputation. When you champion digital trust and privacy, and actually use those tools internally, it sends a clear message that you're an ethical, forward-thinking organization.

Picture this in practice. Your company tells candidates that you'll never ask them to email sensitive information that could be lost or mishandled. Instead, you explain that you use the Trust Credential, which lets them keep control of their data while you verify it securely. That simple moment communicates respect. Anyone who has dealt with invasive, repetitive background checks or insecure HR processes immediately recognizes the difference. Small interactions like these often shape how a company is perceived as an employer.

This same approach scales across the entire gig and on-demand labor economy. Platforms such as Upwork, Instacart, DoorDash, Wonolo, and seasonal staffing apps could make hiring dramatically faster and far less risky by connecting to a shared Trust Credential system and the Trust Bureau. A worker completes biometric identity verification, background checks, and license checks. Their relevant background information is continuously monitored for any updates, and the Trust Credential is automatically updated.

After that, they receive a portable TruaScore from 0 to 360 and a set of verifiable digital credentials they can use anywhere. Employers see only the score and the attributes relevant to the job, never the worker's personal data, which allows them to match people to roles in seconds. This works for long-term freelance projects as well as same-day shifts in restaurants, retail, or delivery.

Onboarding that normally takes days and costs $50 to $200 per hire shrinks to a minimal cost. Fraud, no-shows, theft, and liability issues also drop because the score updates in near real time if new criminal records, sanctions, or licensing problems surface.

For platforms and clients, this means faster talent deployment, lower screening costs, stronger compliance without handling sensitive data, and far fewer bad hires. For workers, a single verified Trua profile becomes a reusable trust passport across every gig app, eliminating repetitive paperwork and giving reliable performers a real advantage.

And the talent effect only compounds. When a business adopts new trust technology, it signals innovation and attracts people who want to work in an environment that's moving forward. The reverse is just as true. Companies that rely on outdated, messy processes often appear careless or behind the curve, which can push strong candidates away. I've had engineers walk away from offers simply because they didn't want their names tied to the next data breach. Trust matters just as much to the people building the products as it does to the people using them.

Inside the company, trust credentials also reduce bureaucracy. Large organizations often require periodic rescreening of employees, especially in regulated industries or when someone is being considered for a promotion. A continuous, portable credential automates most of that work, freeing HR teams from tedious audit cycles. The same benefit applies across the contractor and supply chain ecosystem.

Companies spend huge amounts of time and money managing certificates, proof of training, access badges, and similar documentation. When contractors arrive with a portable credential, onboarding becomes nearly instant. Projects start sooner, delays shrink, and any operation that relies on external labor runs more smoothly.

Benefit 5: Compliance Automation

We touched on this in the previous chapter, but it deserves a deeper look from a business perspective because compliance is one of the most expensive and least scalable parts of modern operations. Companies spend massive amounts on staff hours, software platforms, consultants, insurance, and legal fees just to meet data regulations and complete basic due diligence workflows. None of this work creates new revenue. It is simply the cost of being allowed to operate.

When a company adopts the Trust Credential and Trust Bureau model, many of those costs shrink immediately. If fewer regulations apply to you because individuals maintain control of their own data, the savings can be measured in millions per year. And when your internal systems no longer store or process personal data, your obligations under the GDPR and the CCPA drop sharply. There is no extensive database from which to retrieve, correct, or delete data because the data stays with the user.

This shift also transforms the audit experience. Many industries are required to regularly prove how they conduct background checks, customer data handling procedures, identity verification, and ongoing due diligence. When those steps are handled by a secure Trust Bureau, resulting in you not collecting, storing, or guarding customer data, your business sails through any audit.

The same pattern repeats across sectors such as healthcare and education. Hospitals spend months credentialing doctors and nurses and verifying licenses, certifications, and disciplinary history. Schools and employers invest time and money in validating transcripts and professional qualifications. Any industry weighed down by heavy verification requirements benefits immediately when those processes are simplified and consolidated through a reusable Trust Credential.

And once you step back and look at the broader landscape, the industry-specific opportunities become obvious. Anywhere compliance depends on identity proofing, credential verification, or ongoing monitoring, the Trust Credential and Trust Bureau model removes friction, reduces risk, and cuts costs at a scale traditional systems simply cannot match, all the while dramatically reducing fallout from any breaches.

THE BUSINESS REVOLUTION

So Many Advantages

In financial services, the impact of instant customer verification changes how people feel when they interact with a business. Think about the tension most of us carry when applying for a loan or opening a new account. We hope the process is safe. We hope our information won't be exposed. We hope someone on the other end actually trusts us.

When verification happens instantly through a credential the customer controls, you remove that anxiety. Approval comes faster. Confidence rises. And customers finally feel protected instead of exposed. Even false fraud alerts begin to fade because identity signals are so much stronger. The result is not just efficiency, but goodwill. People return to the places where they feel seen and safe.

A Trust Bureau takes this even further by giving lenders a richer, more human picture of each applicant. Many of us know people who are financially responsible but invisible to the traditional credit system. They've paid rent on time for years or worked steadily but don't have the right mix of credit accounts. With verified income, employment, and payment history available through a trust credential, those individuals finally get the opportunity they deserve. And lenders benefit too, discovering entire groups of reliable customers they could never reach before.

Healthcare becomes more personal as well. Anyone who has filled out endless forms in a waiting room knows how dehumanizing and exhausting that process can feel. Clinics lose valuable time. Patients lose patience. Staff are burning out. And hospitals spend months verifying a doctor's history before they can even start caring for people. A Trust Credential cuts through all that noise. Onboarding becomes smoother. Patients move through the system more comfortably. Doctors and nurses can begin practicing within days instead of months. And because their credentials are verified at the source, everyone in the building can trust the people working beside them. In a field defined by stress, that sense of assurance matters.

The gig economy tells its own story. When you get into an Uber late at night or welcome an Airbnb guest into your home, trust is not an abstract idea. It's visceral. You feel it in your chest. You either feel safe, or you don't. Portable trust credentials strengthen that feeling on both sides. Drivers know their passengers are verified. Homeowners know their guests are legitimate. And the platforms that adopt these tools first become known as the safe places to participate. A simple badge on a user profile becomes more than a symbol. It becomes a promise. People choose the environments where they feel protected.

Government services benefit in a deeply human way too. Millions of people depend on programs such as unemployment benefits, licensing, or healthcare assistance. When verification takes weeks, their lives stall. When fraud occurs, the people who truly need help pay the price. Trust Credentials change that. Approvals speed up. Fraud becomes harder. And citizens feel that the system is finally working with them instead of against them. Businesses that contract with government agencies experience the same shift. Instead of drowning in compliance paperwork for every employee, teams only need to meet a clear trust score. Projects begin faster. Headaches fade.

At the larger level, trust comes down to something deeply human. When a breach happens, it isn't just the company that suffers. It's the people who placed their confidence in that company. A compliance failure doesn't show up as a line item; it shows up as uncertainty and doubt. And a single lawsuit can undo years of steady progress. One incident can drive loyal customers away. Protecting them before anything goes wrong is an act of care, and the Trust Credential and

Trust Bureau model embeds that protection into the core of how your business functions.

There is also a deeper truth many leaders hesitate to face. We are at a turning point. The pressure is rising. The risks are multiplying. Customers are more cautious. Regulators are paying closer attention. People are tired of being told they can trust a company while being asked to surrender data that will not be protected.

This moment asks for a choice. Lead the shift or postpone it and deal with the consequences. Those who take action now will welcome customers more quickly, inspire confidence in every interaction, reduce operational strain, and establish strong working relationships with regulators that value responsible innovation. Those who wait will appear unsafe and outdated, lose business after the next breach, struggle to attract talent, and absorb rising compliance costs that feel heavier each year.

Investment and Change Are Required

Adopting the Trust Credential system requires intention and investment, and facing that level of change can feel intimidating. Some companies will need to adjust login flows, retrain or resize parts of their team, or coordinate with a Trust Bureau provider. Those steps take effort, but they are short-term commitments compared to the long-term gains. The return on investment is substantial. Many organizations can unlock two to five times the value through lower fraud, reduced operational costs, and far more efficient onboarding.

This approach will only work when the private sector leads and individuals maintain control over their own Trust Credentials. That creates a trust ecosystem that serves businesses and consumers without relying on government-issued digital identities. The principle is straightforward: Use an established trust network rather than trying to rebuild one internally, and you benefit immediately from faster processes, cleaner data flows, and fewer losses.

There is also a strategic opportunity for early adopters. Companies that implement Trust Credentials now will help shape the standards others eventually follow. They will be the voices regulators listen to, the ones contributing to industry frameworks, and the ones demonstrating how Trust Credentials strengthen fairness, privacy, and user control. For instance, employers using Trust Credentials in hiring could collaborate with the EEOC to show how verified attributes reduce bias and improve consistency. When guidance evolves in that direction, companies will already be operating in alignment.

This shift toward privacy, data minimization, and user control is gaining momentum. Regulators are increasingly encouraging designs that limit the amount of personal data companies collect and store. Trust Credentials fit naturally into that vision. Adopting them early means fewer compliance surprises later and fewer urgent scrambles to retrofit outdated systems when expectations change.

Some leaders may still worry about implementation complexity. For smaller companies especially, integrating with a Trust Bureau may sound daunting. In reality, the system is built to be simple. Integration is done through clear, modern APIs, much like enabling a payment processor or adding a new login method. It does not require rebuilding your internal infrastructure. The model is flexible enough for organizations of any size.

Concerns about relying on an external Trust Bureau are understandable, yet businesses already depend on credit bureaus, payroll providers, and third-party identity services. The difference here is that Trust Credentials reduce risk rather than increase it. By design, you avoid storing raw personal data, which lowers your exposure and strengthens the protection you offer your customers.

A Shared Trust Infrastructure

Companies must move away from hoarding data and toward an ecosystem where consumers are co-opted to guard their own data. Shared trust infrastructure benefits everyone. If we all rely on the Trust Bureau, trust signals improve across the entire network. This resembles public infrastructure such as roads or the power grid. Commerce becomes smoother when everyone participates.

Building a private trust system within a single company does not create the same value. Network effects matter. A bank may verify a user once, but the true value lies in the ability of that user to present that verification elsewhere. This approach is nothing short of promoting societal good.

Collective trust insights, managed with privacy at the center and consumer agency over their data, produce stronger outcomes for every participant. A helpful analogy is spam filtering. Email providers dramatically improved filtering when they began sharing information about known spam. Had each provider continued to work alone, far more spam would be slipping through. Trust works the same way.

The urgency is real. Regulators are introducing new rules around digital identity and verification. Fines for data misuse are increasing. Consumers are making decisions based on trust and security rather than price alone. Technology such as blockchain and biometric verification has matured enough for large-scale deployment. All the ingredients are in place.

The cost of waiting grows every day. Within five years, companies that do not adopt reusable Trust Credentials will appear as outdated as businesses without websites in the early 2000s. Users will naturally ask whether a company accepts their Trust Credential. If the answer is no, they may take their business or talent elsewhere.

Early adopters will lead the next wave of business transformation. They will set the benchmarks for speed, safety, and customer experience. They will uncover new opportunities made possible by trusted digital identities. Entire markets that once seemed impossible may suddenly be viable. Peer-to-peer insurance, lending, and other trust-dependent sectors could expand rapidly once people can reliably trust strangers through digital credentials. Companies prepared for this shift will grow. Those stuck in manual or siloed verification processes will fall behind.

If you're a business leader, I challenge you to embrace this change now or struggle to catch up later. The Trust Credential and Trust Bureau model aligns the interests of consumers, businesses, and regulators. What helps one helps all. This is rare in technology. By championing trust and privacy, businesses are not merely checking regulatory boxes; they are giving regulators the opportunity to rethink the existence of myriad regulations in the first place. More importantly, it is investing in a stronger and more efficient operation and a more loyal customer base.

Ultimately, this model redefines the economics and architecture of the trust ecosystem. Organizations that proactively master verifiable trust will gain a decisive competitive advantage and deepen customer loyalty. Those who delay will find themselves at a growing disadvantage as trust becomes the primary differentiator in every market.

The path forward is clear: Invest in embracing privacy-centric trust verification systems now, or risk gradual erosion of consumer confidence and relevance. Companies that lead this transformation will capture lasting value; those that hesitate will struggle to regain the trust they once took for granted.

CHAPTER 7

FROM REGULATORY OVERLOAD TO REAL-WORLD RELIEF

This brings us to a topic most organizations would rather avoid: regulation. Businesses operate in a thicket of data regulations intended to protect consumers, yet the outcomes often feel disappointing. Over the years, rules such as the FCRA, CCPA, HIPAA, and GDPR have piled up. Each new law brings additional requirements, audits, and potential penalties. Organizations feel that weight every time they must update a policy, retrain a team member, or pay for another round of compliance software.

While most organizations I talk to are not against protecting consumers, the sheer volume and morass of rules have created a system where you spend more time trying to avoid missteps than building your business. Federal and state agencies publish thousands of pages of

The sheer volume and morass of rules have created a system where you spend more time trying to avoid missteps than building your business.

guidance, and staying compliant can feel like chasing a moving target. It is no surprise that businesses collectively spend billions every year on attorneys, consultants, and training just to keep up. These mandates were meant to help, but in practice, they often slow your team down, create layers of bureaucracy, and make innovation harder than it should be.

And even after all that effort, the results are far from reassuring. Data breaches keep happening, inaccurate records still hurt customers, and public trust continues to erode. You've probably seen it yourself. A customer disputes an error in their file only after the damage has been done, or a background check company sends you outdated information that complicates a hiring decision. Individuals have almost no control over the data that defines them, and that gap is at the heart of many of these problems.

Regulators generally respond by adding even more paperwork, more disclosures, and more complicated dispute processes. In short, despite the steady climb in regulations and compliance spending, trust is still slipping. Businesses feel trapped between protecting their customers and protecting themselves. Consumers feel vulnerable. It is a system that demands more from everyone and delivers less than it promises.

A Prime Example

Take the FTC. Along with the CFPB, the FTC primarily enforces the FCRA, a law meant to keep credit and background reports accurate and fair. In theory, the rules around consent, accuracy checks, and disputes should protect people.

In practice, organizations still end up dealing with the fallout of problems individuals never caused. Because consumers have never had an easy way to inspect or correct their own data up front, errors slip through all the time. Identities get mixed, old negatives linger, and businesses only find out after someone's been denied a job or loan because of information that should have been fixed long before it reached them.

That's when the FTC or CFPB steps in with more rules, fines, and enforcement actions. From the outside, it looks like accountability. On the inside, it feels like whack-a-mole, with businesses absorbing the time and cost of disputes. Their complaint portal now processes millions of grievances every year. In fact, the CFPB received about 1.88 million consumer complaints in just a recent six-month span,[61] which tells you how reactive the entire system has become.

You see the same pattern everywhere else. The FTC can issue major penalties under laws such as the Gramm-Leach-Bliley Act or the Children's Online Privacy Protection Rule if your security or privacy practices fall short, and GDPR regulators can levy fines of up to 4 percent of a company's global revenue for serious data violations.[62] These mandates are meant to prevent abuse, yet most of them activate only after someone's data is already compromised or misused. By then, the damage is done, and the burden is on the individual to prove they followed all the right steps.

Hiring is no easier. The EEOC provides guidance to prevent discrimination, especially regarding the use of criminal records in hiring. But because employers depend on third-party data they cannot verify at the source, they still get false hits, biased reports, and legal

---

61 "Understanding Consumer Financial Complaints: A Deep Dive into Recent CFPB Data," Rahman Legal, March 28, 2025, https://www.rahmanlegal.com/consumer-fraud/understanding-consumer-financial-complaints-a-deep-dive-into-recent-cfpb-data/.

62 "GDPR: Fines / Penalties," Intersoft Consulting, https://gdpr-info.eu/issues/fines-penalties/.

exposure. The EEOC responds by tightening its oversight, leaving businesses with even more to navigate.

The CCPA in California and the GDPR in Europe promise consumers more control. But for businesses, they come with heavy audits, detailed recordkeeping, and system overhauls that cost large companies millions and leave small businesses scrambling. Innovation slows because businesses are constantly looking over their shoulders.

Healthcare tells the same story. The HIPAA requires strict protection of patient records, but the cost of securing systems, training staff, and meeting breach notification standards is significant. Most providers still rely on centralized databases and manual processes that are both expensive and fragile.

It can even delay the rollout of tools such as telemedicine that would otherwise make care easier to access. Cross-border compliance brings another layer of stress as you try to align the GDPR with a patchwork of US state laws, each with its own standards and penalties. You end up maintaining multiple playbooks just to operate across regions.

The result is a frustrating paradox. You are surrounded by rules, responsibilities, and liabilities, yet consumers still suffer data leaks, inaccurate reports, and fading trust. Heavy regulation has not fixed the core problem, because it was never designed to.

As long as personal data is scattered across thousands of databases outside the individual's control, mistakes and abuses are inevitable. Regulators will attempt to create regulations to patch those holes, and you will keep paying the price for a system that was broken from the start.

> *As long as personal data is scattered across thousands of databases outside the individual's control, mistakes and abuses are inevitable.*

A Trust Credential offers a better way forward. Individuals

can store verified health records—such as vaccination history, test results, or insurance status—in a secure, user-controlled credential. Only the necessary data is shared, and only with consent. This reduces the provider's data-handling burden, minimizes compliance exposure, and gives patients true control over their health information. Instead of managing sensitive records across multiple systems, providers simply confirm what has already been verified. The result is lower risk, faster onboarding, and a privacy model that finally matches the expectations of a digital world.

A Paradigm Shift

Here's the good news. By empowering individuals to manage a secure Trust Credential for their identity and history, many of the reactive rules and oversight mechanisms we've accumulated could become unnecessary. Instead of agencies and companies spending resources to clean up data mistakes or chase down breaches after the fact, those mistakes may never happen in the first place.

The Trua Trust Bureau model embodies this shift. Modern cryptographic techniques, such as tokenization and zero-knowledge proofs, enable selective disclosure and let you prove a statement is true without revealing the underlying details. A permissioned blockchain ledger ensures records cannot be tampered with and provides an auditable trail of verifications. Once information is verified and added to your credential, it is reusable and portable across industries and applications.

This model keeps personal data decentralized and user centric. Businesses or agencies no longer need to store large amounts of sensitive information or depend on third-party databases. They simply request the confirmation they need from the individual's Trust Credential via the Trust Bureau. The system answers with a yes or a no, or provides a verified token, without exposing raw personal data.

REGULATORY OVERLOAD TO REAL-WORLD RELIEF

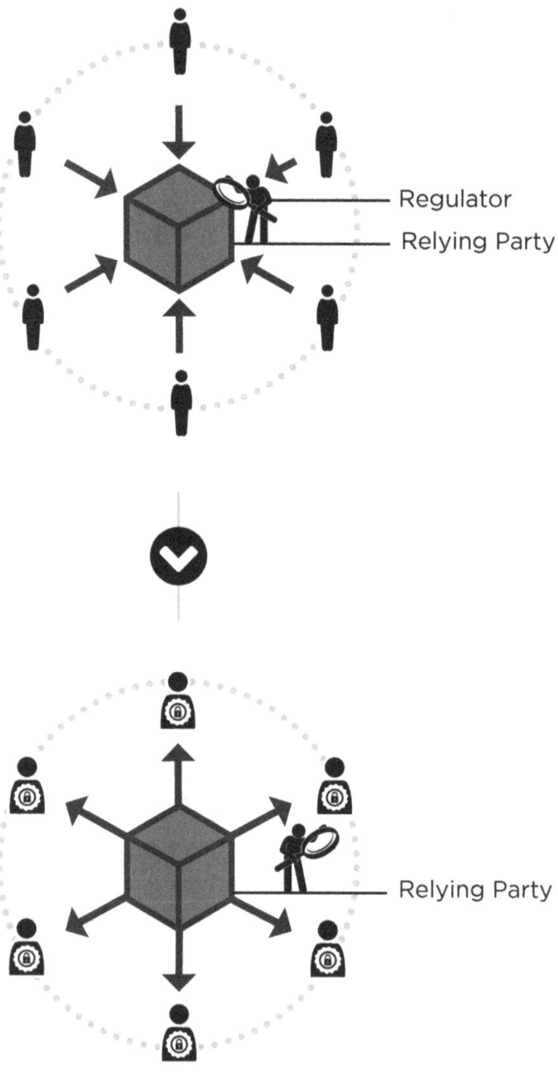

Because verification comes directly from the user-controlled credential and relies on tokenized answers that reveal only the minimal facts required, companies gain the assurance they need without ever touching the underlying data. This sharply reduces the risk of breaches or misuse. It also slashes liability and compliance burdens, because companies are no longer warehousing sensitive personal information. There is less for hackers to steal, and fewer laws and audits to manage. Trust is established up front through technology and user consent, not through a maze of reactive regulations and penalties.

Think of it like a Visa credit card transaction. When you tap your card at a store or make an online purchase, the merchant does not receive your full bank account details or personal identity. They get a secure authorization token that simply says, "Yes, this payment is approved." The underlying financial network confirms the card you presented, and available balance, but the merchant never needs to see your income, credit score, or address to complete the sale.

The Trust Credential model works in the same way. A business does not need your SSN, employment history, or full background report to move forward. It just queries your Trust Credential for exactly what it needs, such as confirming that your license is active or that you passed a background check within the last year. It then receives a verified yes or no response, without gaining access to the raw data itself.

This keeps sensitive information in the hands of the user, not scattered across corporate servers. It also reduces delays, improves security, and, by design, ensures compliance with privacy laws. Just as tokenized payments increased consumer confidence and reduced fraud in digital commerce, Trust Credentials can raise the standard for digital identity, access, and verification across industries. When trust is built into every interaction, everyone benefits.

This shift has significant regulatory implications. The Trust Bureau uses a first-party data model and, therefore, often does not fall under laws such as the FCRA, which were built around third-party credit bureaus. The FCRA defines a *consumer reporting agency* as any person that, for fees or on a nonprofit basis, regularly engages in assembling or evaluating consumer information for the purpose of furnishing reports to third parties.[63]

Trua's system avoids that definition because it does not aggregate or sell consumer reports to third parties. Individuals opt in, obtain their verified Trust Credential, and then share their own verified data directly with those who need it, under their own control. Companies using the Trust Bureau can bypass many FCRA obligations, such as formal dispute handling and adverse action notices, because no third-party credit report is involved in these exchanges.

If background information comes straight from the consumer's Trust Credential rather than from a reporting agency, it likely falls outside the FCRA's scope. This is a major shift. Businesses can gain reliable, verified information without triggering the costly compliance machinery of a traditional consumer reporting process. Early results show that this approach also improves accuracy and fairness: Users can correct and update their own records, catching errors that might have lingered in a third-party database, and only the necessary facts are shared for each decision. In other words, accuracy is improved at the source rather than enforced later through disputes and lawsuits.

By addressing root issues and bringing individuals into its data review and verification loop, the Trust Bureau model can achieve regulators' consumer protection goals with far less complexity. It aligns with policymakers' calls to simplify regulatory systems, showing that consumer empowerment can support rather than conflict with deregulation. Instead of adding more rules every time something goes wrong, we can redesign the system so it naturally prevents those issues.

---

63 "15 U.S. Code § 1681a - Definitions; Rules of Construction," Legal Information Institute, Cornell Law School, https://www.law.cornell.edu/uscode/text/15/1681a.

It's like moving from an environment with endless safety nets (with some big holes, I might add) to one where the tightrope itself is safer to walk. This model reduces the widespread duplication of personal data. For instance, freelancers or gig workers no longer have to scatter their SSNs and ID scans across multiple platforms, which in turn lowers identity theft risk.

Hiring, Employment, and Onboarding

Hiring provides a clear example of regulatory overload versus real-world relief. Today's hiring processes often involve extensive background checks governed by laws such as the FCRA and guided by EEOC best practices.

Traditional screening can take weeks, as third-party agencies pull records from databases and public sources. It is not only slow but also prone to error. Outdated or incorrect records can disqualify good candidates, and biases can creep in when human reviewers interpret the data. Employers also carry the burden of handling applicants' sensitive information, including SSNs and detailed personal histories, with the constant risk of data breaches or legal exposure if anything goes wrong.

In most hiring systems today, employers are allowed to conduct a formal background check only after they extend a conditional job offer. This requirement exists because the background check is performed by a third party, which brings legal obligations under the FCRA. As a result, employers are forced to make an offer before verifying critical information. If issues arise later, the employer may have to revoke the offer, which introduces friction, risk, and, in some cases, legal consequences.

The Trust Credential model changes that. Instead of waiting until the final stage to verify someone's background, the individual can complete verification independently. They confirm their own education, licenses, and court records, then hold that data in a secure, user-controlled Trust Credential. When applying for a job, they choose what to share and with whom. The employer can receive accurate, verified information at any time during the hiring process without having to request or store sensitive personal data, and without triggering regulatory complexity.

This model improves the process for both sides. It is faster, more secure, and avoids unnecessary delays. It lets candidates present their qualifications confidently and lets employers choose a candidate with clarity, well before the offer stage.

With a user-controlled Trust Credential, much of this friction disappears. Job seekers verify their own credentials once, then share proof of specific qualifications or a clean record directly with employers. An applicant can instantly provide confirmation of their education and the absence of a criminal record without the employer ever seeing full background details or storing any sensitive data.

Picture this example. A software firm is overwhelmed with résumés, and a candidate submits her Trust Credential. It includes a verified degree, professional certifications, and proof of identity. The recruiter receives a clean, simple dashboard showing that this candidate's degree was confirmed, her certifications are current, and no disqualifying criminal record is on file—everything needed to move forward.

It arrives instantly, bypassing weeks of manual background checks. Because the Trust Credential is maintained by the user herself, it is always up to date and has already been verified for accuracy via the Trust Bureau.

When the hiring manager wants to confirm any details, they do not initiate a new screening process. Instead, they query the Trust Bureau directly, receiving a cryptographically verified confirmation that the information is legitimate. No raw data is exchanged. No paperwork is needed. The process is fast, secure, and anchored in user consent and privacy.

This not only speeds up the hiring process but also reduces the chance of bias. The hiring manager sees only the relevant validated facts, not extraneous personal details that might trigger unconscious prejudice, such as zip codes or addresses. The impact on efficiency and fairness can be significant.

By letting individuals control and present their information, the process becomes more focused on actual competencies and less on bureaucratic hurdles. Businesses fill roles faster with vetted hires, and candidates are judged on their real qualifications.

Compliance becomes radically simpler. In this model, employers never handle protected personal data directly, which slashes their exposure to privacy regulations and the legal liability that comes with it. They're not functioning as a "consumer reporting agency" under laws such as the FCRA; they're simply receiving a user-approved, verified Trust Credential. That one distinction dramatically reduces employers' compliance burden, freeing them to focus on hiring the right people instead of wading through bot résumés, legal disclosures, dispute protocols, and audit trails.

Just as important, this first-party model curbs bias and error at the source. Candidates get the opportunity to verify and update their information before it's ever shared. Outdated or incorrect records don't quietly sabotage someone's chance at employment. Employers get accurate, relevant data—no noise, no guesswork. The result? A hiring process that's faster, fairer, and more transparent from the start.

Financial services face a similar choke point. Institutions are required to verify customer identities through KYC checks before someone can open an account, apply for a loan, or move money.

For many users, especially those outside the credit mainstream, this becomes a frustratingly repetitive and exclusionary process.

A Trust Credential flips that dynamic. It allows individuals to prove their identity and credibility directly, with full consent and without handing over piles of raw documents. That streamlines both compliance and security. Financial institutions gain trustworthy, cryptographically secured signals—without storing sensitive personal data themselves. Customers who were once sidelined by rigid identity verification processes and rules now have a portable way to demonstrate their legitimacy and credibility. And financial firms reduce their fraud risk and compliance overheads in the process.

The trust isn't in the paperwork anymore. It's in the Trust Credential and the math that backs it.

Healthcare and Privacy

Healthcare is another arena where strict regulations meet daily friction. Privacy laws such as the HIPAA demand extreme care in handling patient information, which helps preserve trust—but often at the cost of speed and flexibility. Before a telemedicine consultation, for instance, clinics must verify a patient's identity and gather their medical history in privacy-compliant ways. That typically means paperwork, repeated identity checks, and a frustrating back-and-forth for both patients and providers.

But it's not just patients who need to be verified. In a rapidly growing telehealth landscape, healthcare providers themselves must also be screened and credentialed—especially when they are engaging with vulnerable patients. Whether it's a virtual therapist, an urgent care doctor, or a nurse sent to a patient's home, patients need to know that the person on the other end is legitimate, licensed, and in good standing. Without that assurance, trust breaks down fast—and with it, the willingness to seek care remotely.

The Trust Credential model addresses both sides of this equation. It maintains privacy and security without the usual administrative drag. Patients can carry verified health credentials in a digital wallet, including proofs of identity, age, and insurance coverage or tokenized snippets of medical history such as a vaccination record or lab result. They share only what's needed with a healthcare provider.

Meanwhile, providers themselves can be verified with a similar credential, confirming licensure, background, and medical authority in real time. A patient scheduling a telehealth visit could instantly confirm that their physician is board certified and authorized to practice in their state—no guesswork, no blind trust. The system cuts through bureaucracy while boosting confidence, creating a healthcare environment that's safer, faster, and more equitable for everyone involved.

A telehealth platform can then confirm a patient's identity and key medical facts without ever receiving full medical records. One example from 2025 involved an elderly patient in a rural area using a telemedicine app. Before her appointment, she shared her credential, which showed that her identity had been verified through a biometric match and included a token confirming a recent blood test result—no other medical details exposed. The specialist proceeded confidently with the vital information already confirmed.

A study by a telehealth provider showed that this method made remote consultations faster on average, and one telemedicine platform saw diagnosis times improve by about 40 percent thanks to streamlined digital verification.[64] Privacy-related issues and frustrations

64 Vinati Kamani, "Store and Forward Telehealth: 2025 Implementation Guide," Arkenea, August 10, 2025, https://arkenea.com/blog/the-complete-guide-to-store-and-forward-telehealth/.

dropped as well. When patients feel more secure and in control of their data sharing, they are less likely to lodge privacy complaints.

In fact, introducing easier identity checks has tangible benefits. For example, one report found that AI-enabled eligibility verification reduced patient no-shows by 35 percent, indicating smoother, more trusted appointment setups.[65] Doctors end up spending less time on bureaucratic identity confirmation and more time on care, while patients avoid filling out repetitive forms and the worry that their entire medical history is being transmitted or stored unnecessarily.

Healthcare professionals benefit in other ways too. Consider traveling nurses or locum tenens doctors who move between hospitals and states. Each new assignment traditionally requires reverifying licenses, certifications, vaccinations, and background checks, a process that can take days or weeks and creates delays when staffing is urgently needed.

With a portable Trust Credential, a nurse can keep all qualifications and clearances pre-verified in their profile, updating it whenever they gain a new credential. When a crisis arises in another state, they can instantly share their up-to-date Trust Credential, showing, for example, that their RN license is active in that state, their specialty certification is valid, and their immunizations are current.

Smart contracts could even automatically cross-check these against licensing databases. A pilot program showed that this approach cut emergency placement times significantly. One hospital network reported onboarding traveling nurses about 65 percent faster during a staffing shortage. Even more conservative implementations see major gains; for instance, centralized credential verification systems have been shown to reduce hospital nurse onboarding times by roughly

---

65 Ibid.

40 percent.[66] Faster onboarding meant help reached patients sooner, without compromising compliance. Hospitals didn't have to sift through faxed paperwork or juggle different state requirements. The verified credential did the heavy lifting.

The benefits extend to everyday healthcare operations. Pharmacies, for example, can fill online prescriptions knowing the credential confirms the prescriber is licensed, the prescription is valid, and the patient is over eighteen—all without the pharmacy ever seeing full identity details or medical files.

Stuck in Reactive Mode

Caught between inconsistent laws and aggressive enforcers, many organizations have been playing a losing game of whack-a-mole. They scramble to plug compliance gaps only after a new law hits or a regulator comes knocking. It's a reactive, defensive posture that isn't working. The more companies chase each new rule or settlement decree, the further they fall behind. Compliance budgets skyrocket, yet confidence in actually being compliant plummets. Business leaders are understandably frustrated, even exasperated, by this cycle of ever-expanding obligations and gotcha enforcement.

It doesn't have to be this way. The Trust Credential and Trust Bureau model offers a path out of this reactive cycle and into a proactive stance. Instead of trying to comply with a dozen different data-handling rules in a dozen jurisdictions, imagine flipping the script on data privacy: Give consumers control of their own data and verify it at the source.

---

66 Sarah Knight, "Streamline Nurse Onboarding: How Centralized Credential Verification in Healthcare Cuts Hiring Time," ShiftMed, November 17, 2025, https://www.shiftmed.com/insights/knowledge-center/leveraging-centralized-credential-verification-for-healthcare-workforce/.

The benefits are game-changing. No unverified personal data is shared, and there's no need for companies to stockpile sensitive information on their servers. The consumer's data is vetted and accurate (because the consumer themself confirmed it), and it's only disclosed to the company with the consumer's explicit permission. In essence, this approach "reroutes control over consumer data, placing it in the hands of the consumer."

By design, a trust-based, consumer-centric data ecosystem sidesteps many of the thorniest compliance issues. If a business isn't collecting piles of unsolicited personal data, it doesn't have to worry about a patchwork of state data-selling opt-out rules because there's nothing to sell or leak inadvertently. If an individual directly provides a verified credential (for, say, their background check or creditworthiness), the company can be confident the information is accurate and FCRA compliant, rather than praying that some data broker's report doesn't land them in legal hot water.

In short, shifting to a first-party, consumer-verified and -consented data model transforms privacy compliance from a relentless burden to a built-in feature. Instead of constantly reacting to regulators with apologies and retrofits, organizations can finally get ahead of the game. The Trust Credential and Trust Bureau framework is a chance to streamline the thorny issues of data gathering and use, turning a regulatory quagmire into a competitive advantage. It's time to break the cycle of reaction and embrace a model that satisfies regulators and customers alike—by design, not by after-the-fact Band-Aids.

Gig Economy Platforms

The gig economy faces a serious trust problem. Millions of transactions happen daily between strangers—drivers and passengers, caregivers and families, freelancers and clients—on platforms that often verify little beyond a scanned ID. Ratings and reviews help, but they are easy to manipulate, rarely updated, and offer no assurance that a person is who they say they are or that their qualifications are current.

Most platforms today rely on fragmented, one-time background checks, often done by third parties. The process is slow, inconsistent, and quickly becomes outdated. Workers have to repeat the same steps for every new app. Customers are left guessing whether the person on the other end has been properly screened, if at all.

The Trust Credential model changes this. It gives platforms the ability to continuously verify both sides: service providers and service seekers. A rideshare driver can present a current, verified token confirming a clean driving record and license status. A customer hiring a home caregiver can show verified ID and payment credentials, reducing fraud and increasing accountability. Trust becomes mutual and real-time.

Outdated checks are replaced with live credentials that update automatically. If something changes in a worker's status, the platform sees it immediately. This makes it far harder for bad actors to slip through and keeps the trust level high over time. It also reduces liability and removes the burden of repeat vetting from both the worker and the platform.

This is not just about convenience or safety as nice-to-haves. When platforms get trust wrong, people get hurt. Fraud, theft, or worse becomes a real possibility. Consumers expect more, and regulators are starting to demand more.

The Trust Bureau model answers that call. It turns trust into a shared, portable credential that workers can carry across every app. It replaces guesswork with proof. It gives customers a reason to feel safe and platforms a smarter, cheaper way to grow. In a market built on strangers, real-time, verified trust is no longer optional. It is the baseline for what comes next.

Government Services and Civic Applications

Government programs and civic services show the same contrast between regulatory overload and real-world relief. Take welfare and public benefits systems as a clear example. Verifying eligibility for unemployment or food assistance often requires applicants to submit stacks of documents such as pay stubs, IDs, utility bills to prove residence, and so on.

Agencies must follow strict verification protocols to prevent fraud. The process is slow, error prone, and stressful. People end up submitting repetitive paperwork that exposes a lot of sensitive details, and despite all the rules, some fraudsters still game the system, while some eligible individuals fall through the cracks due to delays or missing documentation.

A Trust Credential can streamline these interactions enormously. Instead of repeatedly sending sensitive documents to different offices, applicants share digital proofs of the required criteria directly from their Trust Credential. For example, someone applying for unemployment benefits could prove they earned a specific amount in the last quarter and that they were laid off from a particular company by sharing tokens tied to verified tax and employment records.

This approach also protects privacy and autonomy. Applicants share only what is needed to prove eligibility, not a thick file of personal documents that will sit in an archive. Because the data originates under the individual's control, it reduces the sense that agencies are prying into private lives. Administrators benefit too.

They spend less time chasing down documents or dealing with identity disputes and more time actually helping people. For instance, a single parent applying for childcare assistance could verify their income and employment in minutes through their Trust Credential, receive aid faster, and avoid the stress of multiple office visits or the exposure of their personal info. Public resources are used more efficiently with fewer bureaucratic steps and the same or better outcomes.

Even core civic duties, such as voting, can benefit from this model. Election systems must protect integrity while remaining accessible. Instead of relying only on centralized voter rolls and varying voter identification laws, a Trust Credential could allow voters to prove their eligibility by presenting a verified token at the polling station. The token could simply confirm "yes, this person is a registered voter in precinct X" without revealing their name or address to the device, thus protecting privacy.

Biometric liveness checks could ensure the voter is real and present. And a blockchain record could note that the credential was used to vote, preventing anyone from voting twice without storing personal data centrally. Such approaches can reduce disputes over eligibility and potentially increase turnout by making voting more convenient and secure.

It especially helps those who have difficulty reaching a polling station. For example, overseas military voters or people with disabilities could vote safely and remotely by enabling safe remote verification of their ballots. This method essentially adds trust to the electoral process by providing a cryptographically assured way to verify voters, reducing the need for some of the more burdensome voter identification paperwork and manual checks that have become contentious and error prone.

From welfare offices to polling stations, shifting verification to a user-centric, cryptographically secured model reduces the need for many prescriptive regulations and layers of oversight. Integrity is built into the system itself. Agencies can meet their mandates with less intrusion and red tape. Citizens experience smoother interactions, faster service, and greater control over their information, which can help rebuild trust in public institutions, a crucial goal in an era of skepticism.

Toward Real Relief and Trust

Let's not sugarcoat the reality. Most organizations have not been proactive about earning trust. They have been reactive, doing the minimum required to stay out of trouble. And who can blame them? When regulators speak in riddles and the legal risk of innovation is higher than the risk of doing nothing, the natural instinct is to play it safe. So companies lawyer up, outsource responsibility to vendors, hide behind disclosures no one reads, and look for ways to work around the intent of the rules without actually fixing the root of the problem.

On one hand, I have a 20 percent level of sympathy for business owners. Compliance has become overwhelming. Every new law brings new documentation, new training requirements, new audit risks. If you are a small or mid-sized business, you are stuck choosing between product development and another privacy review. If you are a large enterprise, you are spending millions to build the appearance of control while knowing full well that one third-party breach could unravel everything.

But it's also undeniable to me that many companies have taken the easy way out for far too long. They have leaned on gray areas and legal nuance to avoid real change. They have invested more in navigating loopholes than in building trustworthy systems. They have treated regulation as an external burden to manage rather than a signal that something deeper needs to be rebuilt.

It's time to flip that mindset. If the system is broken, then continuing to play the game by its old rules is not strategy; it's surrender. You cannot compliance your way into trust. You cannot hide behind third parties forever. At some point, you must take real ownership of how your business handles personal data.

If the system is broken, then continuing to play the game by its old rules is not strategy; it's surrender. You cannot compliance your way into trust.

There is a better model, and it is not hypothetical or years away. When individuals hold and verify their own data, and businesses simply request proof of specific facts with the user's consent, the entire calculus changes. You no longer have to collect or store sensitive personal information. You no longer have to guess whether your data is accurate or whether it came from a source you can trust. You no longer carry the same legal and operational risk, because you are no longer in the business of managing data you don't need.

Compliance becomes less about defense and more about design. Trust becomes embedded in the process itself. Consumers feel safer because they are in control. Businesses move faster because they are not bogged down in endless validation cycles. Regulators see less risk and spend less time chasing violations. Everyone wins—not through loopholes or delay tactics, but through structural change.

Sidestepping Legacy Liabilities

The pressure isn't just operational. It's legal, and it's growing. Nowhere is this more obvious than with the FCRA. For any company relying on third-party data to make decisions about individuals—whether that's hiring, lending, or tenant screening—the law imposes strict requirements and steep penalties. The rules were written to protect consumers, but in practice, they often punish the businesses trying to function inside an outdated model.

When you use background checks or credit reports pulled from third-party agencies, you're exposed to risks you cannot fully control. If the data is wrong, outdated, or collected without proper consent, your organization is still the one held liable. You're required to notify applicants, manage disputes, track corrections, and absorb the legal risk if anything slips through the cracks. It's not just burdensome. It's constant legal exposure you didn't create but are expected to manage.

Contrast that with a model where the individual owns their data and shares it directly. A job applicant verifies their background once, stores it in a Trust Credential, and provides it to you at the right moment. A renter verifies their income and employment status on their own terms, not through a third-party database with unknown sources. This is not theory. This is already happening.

When the data comes straight from the user and is pre-verified through a secure and trusted channel, you're not dealing with a consumer report under the FCRA. You're receiving a piece of information the individual chose to share—accurate, up to date, and confirmed. That single shift drastically reduces your exposure. You avoid the procedural traps of legacy compliance models. You get the clarity you need without the liability that used to come with it.

And this goes well beyond the FCRA. The broader data ecosystem—data brokers, social media scraping, shadow profiles—is becoming legally radioactive. Regulators are cracking down. Enforcement actions are rising. Laws are tightening. The tolerance for companies using personal data without clear consent or transparency is evaporating.

If your current model still relies on harvesting or buying third-party data without individual control at the center, it's not just inefficient. It's dangerous. You're building on legal quicksand.

The Trust Credential model doesn't just sidestep these risks. It aligns with where the law is going. It meets the growing demand for consent, transparency, and user control. It reduces your legal overhead and strengthens your position with regulators. And it gives your customers a reason to believe you're doing the right thing, because they see the process, control it, and benefit from it directly.

Aligning with Emerging Privacy Principles

Shifting to a consumer-first, first-party data model does more than keep you in regulatory compliance. It changes the posture of your entire business. Instead of bracing for the next privacy requirement, you move in step with the principles driving this global shift—privacy by design, accountability at the edge, and trust as an active ingredient in every transaction.

When individuals verify and share their own data directly, your organization stops carrying the legal baggage of overcollection, questionable sourcing, and outdated records. You no longer have to justify how you got the information or prove that it was gathered under the right terms. The audit trail begins with the user, and the scope of what you handle is limited to what's actually needed.

This approach also helps close the gap between what the law requires and what users expect. People don't want hidden permissions or vague notices. They want to know what they're sharing, who's seeing it, and what happens next. A Trust Credential offers that clarity by default. Every request is intentional. Every disclosure is purposeful. You don't need to overhaul your privacy language every time laws change, because the foundation is already sound: Consent is embedded, scope is limited, and visibility is built in.

And when the next wave of regulations does arrive, you're not scrambling. You've already replaced passive compliance with active governance. You've reduced the attack surface, simplified your policies, and cut your dependency on third-party data you never truly controlled. You're operating with precision, not guesswork.

In a market where trust and usability are converging, companies that adopt a consumer-first model gain more than just regulatory insulation. They gain credibility. They remove friction. And they create systems that scale without compounding risk.

Making this shift is not about preparing for what might happen. It's about recognizing what already has. Privacy is no longer a legal department issue. It is a product decision, a brand signal, and a competitive differentiator. Moving to a model where users manage their own data is not just the safer path forward; it is quickly becoming the only one that makes sense.

Get Ahead of the Curve or Get Left Behind

Businesses that move early on first-party, consumer-controlled data aren't just preparing for what's coming down the road; they're shaping it. They are making a strategic choice to operate on their terms, not on regulators' timelines or crisis management cycles.

While others wait for mandates to tighten or headlines to force action, early adopters are already reducing legal exposure, simplifying compliance, and earning real trust.

A company that puts user control at the center of its data practices sends a clear message that they respect their customers enough to let them decide what to share. That kind of clarity builds market credibility. It also aligns with the growing expectations of lawmakers, business partners, and consumers alike.

Companies that embrace user-first frameworks now are influencing the standards, not scrambling to meet them. They are better positioned to guide industry dialogue, set expectations, and avoid the reputational damage that comes from lagging behind on privacy and transparency.

There are few decisions in business where legal risk and competitive differentiation point in the same direction, but this is one of them. By adopting a trust-based, user-driven model now, companies reduce complexity, improve credibility, and move faster with less friction. They avoid the firefighting and stand apart from peers who are still hedging.

CHAPTER 8

YOUR ACTION PLAN

Despite all the ground we've covered in this book, you might still be thinking there is little good that you can do. After all, you're just one person, and the digital avalanche seems so large. But I want to assure you, your voice matters!

History shows that meaningful change almost always begins with a single decision, a single shift in awareness, a single moment when someone chooses courage over complacency. That spark has ignited revolutions, corrected injustices, reshaped industries, and transformed nations. In this journey, you are that spark.

You have seen how the digital world has been working against you, treating your personal information as a commodity to be collected, traded, and exploited. You have felt the frustration of breaches, scams, and constant manipulation. Now you understand the deeper forces behind it, and you hold the knowledge that can help change everything.

You do not need perfect conditions to begin.

But despite what you might think, you do not need perfect conditions to begin. You do not need to understand every technical detail. You do not need to wait for someone else to fix what is broken. You can start now, with what you have.

If You're a Consumer

As a consumer, you might not think of yourself as powerful, and it's easy to feel like you are simply using apps, shopping online, getting a job or a gig, and trying to live your life. Yet you hold immense influence through your choices and your voice. Believe me. Companies fight for your trust and your dollars, and policymakers ultimately answer to your voice and vote.

The key is to recognize this. Rebuilding digital trust begins when individuals like yourself refuse to be complacent. It begins when you take ownership of your digital life and insist that companies and institutions treat your information with respect.

Your first responsibility is to protect yourself. Starting today, I challenge you to strengthen your digital habits and make sure the basics are solid. Enable two-factor authentication on important accounts so that a code on your phone prevents automated attacks. Use a password manager so each password is unique and strong. Freeze your credit if you are not opening new accounts. Review your privacy settings and remove permissions that apps do not need.

This week, begin reclaiming any personal information already scattered across the internet. You might be stunned to learn how many data brokers and companies hold pieces of your identity. Start a data cleanse by requesting reports from major data brokers and opting out wherever possible.

If you have done business with companies that suffered breaches, and most of us have, *assume* your information was exposed. Update your passwords, change your security questions, and delete accounts you no longer use. Because old profiles and abandoned accounts act as unlocked doors into your life, do whatever it takes to close them. This is also the time to update your devices and install reputable security tools. Treat your personal data as you would treat your money and guard it with intention.

As the month unfolds, begin using your influence as a customer. Examine the apps, websites, and services you rely on and ask whether they deserve your trust. Look at how much data they collect and how they verify your identity. If they still demand sensitive information without offering modern and safer alternatives, say something.

Companies do listen when enough people speak, so use your voice to ask whether they plan to support privacy-first verification or reusable Trust Credentials. Then back your words with action. Support businesses that prioritize your safety. You hold that same power, so use it.

In the coming year, expand your role from protector to advocate. Share what you have learned with those around you. Many people already sense that something is wrong but do not know where to begin. Explain that there is a better, safer way to prove who we are online without exposing sensitive information. Encourage your community to take similar steps. Support legislation that protects data rights. When petitions and calls for public input appear, add your name and consider joining local digital rights groups. Community centers and libraries often host digital literacy workshops, and you can volunteer to help teach others.

On a personal note, stay alert to developments around the Trust Credential and the Trust Bureau. Pilot programs will appear, and early adopters shape the path forward. When a trusted digital credential becomes available, sign up and begin using it. Then let companies know that if they want your business, they must offer verification methods that do not expose your personal information. Just as early users of TSA PreCheck transformed airport security by proving there was a better model, early adopters of Trust Credentials will do the same for digital life in the AI era.

Above all, stay vigilant. Never assume someone else will fix the problem for you. If you ever feel like your voice does not matter, remember that consumer pressure works. Companies have rewritten policies because customers demanded it, and lawmakers have advanced legislation because citizens refused to be silent.

So use your voice. Ask why a company needs your information, ask how they protect it, and ask whether there is a safer alternative. Each question reminds the powerful that trust must be earned. And each question reminds you that you play a central role in rebuilding the digital world. If consumers like yourself demand a Trust Credential world, that world will come.

If consumers like yourself demand a Trust Credential world, that world will come.

If You're a Business Leader

If you are a business leader or entrepreneur reading this, consider yourself both a key ally in this mission and, to be blunt, part of the reason it is needed. As a leader, you have the power and the *obligation* (I want to emphasize this word) to change how your company handles trust and privacy. Customer trust is as tangible as your product inventory. Lose it, and you lose everything.

While it's often said that guilt isn't a good motivator, in this case, I believe it is. If you have, in any way—intentionally or unintentionally—been complicit in mishandling your clients' data, let this book serve as a strong wake-up call. Let the discomfort you feel push you to become a stronger leader and better business owner.

If you have not taken this issue seriously, this is the moment to begin. There is significant upside. By leading with trust, you gain a competitive advantage. You position your brand as one that respects consumers rather than exploits them.

Start by adopting a privacy-first mindset. Shift away from the idea that data is an asset to stockpile. Understand that personal information is a liability unless it is essential. Audit everything you collect. And the moment you uncover data that you do not need, stop gathering it.

If you do not need someone's middle name, SSN, or a high-resolution ID scan, do not ask for it. You cannot lose what you do not store. Data minimization sits at the heart of the Trust Credential model. You verify what you need to know, but you stop hoarding raw identifiers.

Reducing your data footprint reduces your exposure dramatically. When companies tokenize information and store reference codes instead of actual data, breaches become nonevents. There is nothing valuable for criminals to steal. That means fewer panicked calls to legal teams, fewer expensive notification letters, and much stronger customer trust. Make the question *Do we really need this data?* a constant part of every meeting.

You cannot lose what you do not store.

Then invest in the systems that support trust. Strengthen your cybersecurity so that encryption, access control, and monitoring are baseline expectations rather than afterthoughts. Go further by integrating tools that allow for reusable digital credentials.

Instead of keeping background check files or storing copies of IDs, rely on trusted exchanges that tell you only what you need to know. You get confirmation that someone is eligible for hire or that a license is valid without ever seeing the underlying personal information. This shift protects the individual and protects your company. Train your teams so that *No PII by design* becomes part of the culture. Reward product managers and departments who reduce data storage and simplify trust processes. Aim to become a company that holds almost no sensitive personal data. When your employees can proudly say that your business stores minimal PII, you will know your culture has changed.

Embrace the Trust Credential and join the trust ecosystem early. When it comes out, pilot it within parts of your organization. If you run a platform, allow users to verify their profiles through a trusted credential provider rather than uploading their documents. If you hire at scale, begin accepting reusable Trust Credentials that applicants can carry from job to job. By partnering early with a Trust Bureau, you help shape the standards that will define the next decade. You demonstrate leadership in an area where most companies are still behind.

Also examine where you can remove raw identifiers from your internal workflows. Modern technology allows you to analyze behavior and trends without tethering the data to specific individuals. Use privacy-enhancing methods so you can understand your customers without surveilling them. Show the world that a company can grow, compete, and innovate without spying on every click and keystroke.

Transparency is your next responsibility. Trust is not only built in the background but outwardly communicated. Rewrite your privacy messaging so it is clear and honest. Promise that you will collect only what you need and protect it as if it were your own. Then prove it through certifications and regular security testing.

If something goes wrong, admit it quickly and fix it. Customers forgive mistakes when leaders take responsibility. They do not forgive secrecy or excuses. If you support Trust Credentials, advertise that proudly. Many consumers will choose you for that reason alone.

Finally, lead in your industry. Trust cannot be rebuilt by one company alone. Work with your associations to set higher standards. Share what is working for you. Break the myth that privacy is too costly or too complex. Speak publicly and boldly about why trust matters. Collaborate with competitors to adopt common verification frameworks.

Recognize each other's Trust Credentials and reduce the number of places where personal data must be stored. Push for interoperability, encourage innovation, and build metrics that allow you to track trust within your organization. Fewer incidents, improved retention, and stronger customer satisfaction are proof that you are heading in the right direction. Trust can be measured, so use those measurements to fuel consistent improvement.

As you take these steps, you are helping build the future marketplace. In this journey, you play the role of the strategist who provides the infrastructure that makes safety possible. It takes courage to change habits, confront past mistakes, and invest in new approaches. But remember why you entered business in the first place. You wanted to solve problems, sell products and services to consumers, and contribute something meaningful.

The trust crisis is a problem that demands your leadership now. By stepping up, you protect your company, your customers, and the broader digital world. Feel the discomfort, feel the guilt, and let both drive you forward.

If You're a Policymaker

It's helpful to recall that when the internet was first developed by DARPA, the identity layer was left out. That missing layer has come at a steep cost. Society is now paying the price in the form of widespread fraud, fragmented verification systems, and regulatory sprawl. Government agencies have expanded to try to manage the fallout, yet the regulatory framework has become so complex that it is now difficult to navigate, let alone enforce consistently. The time has come for policymakers to address this foundational gap and create conditions that allow both innovation and consumer protection to thrive in the AI age.

If you are a policymaker, regulator, or lawmaker, you carry a unique and urgent responsibility. Consumers and businesses can drive change, but true structural reform requires legislative and regulatory leadership. For too long, policy has lagged behind as technology has leapt ahead. Most data protection laws and identity systems were built for a different era. Meanwhile, bad actors operate across borders with sophisticated tools and little resistance. The current legal framework is not equipped for that reality. It needs to be rebuilt with modern capabilities in mind.

You have the authority to shape the environment that allows a trust-based digital ecosystem to emerge. By updating outdated statutes and removing legacy barriers, you can open the door for models such as the Trust Credential and Trust Bureau to gain traction. You can hold organizations accountable for building trust responsibly, and you can do so in ways that reinforce public confidence across political lines. Privacy, security, and personal agency are not partisan values. They are shared concerns that cut across ideologies.

> *You have the authority to shape the environment that allows a trust-based digital ecosystem to emerge.*

A key step is to modernize the laws around identity and verification. Many of the rules in place today were designed for a world of static background checks and centralized data brokers. These rules rely on disclosures and paper-based workflows that made sense when consumer information was pulled from behind closed doors. That model no longer fits. In contrast, the Trust Bureau gives people control of their data, keeps it accurate and current, and allows them to share it when and how they choose.

Policymakers should make clear that first-party, consumer-driven credentials are fundamentally different from traditional credit reports and should not be regulated in the same way. Doing so may require new legislation or regulatory guidance that supports ongoing, consent-based verification. The goal should be to protect consumers without stifling the very innovations that can eliminate the risks regulators have been trying to contain for decades.

To do this right, collaborate with experts in privacy, cybersecurity, and digital identity. Update the legal framework to reflect modern tools such as tokenization, encryption, and verified consent. Establish a national Trust Credential standard that can be used for employment, licensing, or age verification, just like a driver's license from one state is accepted as valid in all others. When government recognizes a new framework for trust verification, the private sector quickly follows.

Next, you must support the creation of a national Trust Bureau or enable its establishment. This model is a hybrid and consumer-centric alternative to the traditional credit bureau system that has shaped identity management for decades. Policymakers can pursue a public or public-private approach or create legislation that allows private entities to operate Trust Bureaus under strict guidelines. Imagine a Trust Bureau of America that operates to serve the public good, much like national credit unions or public utilities. Regulators should strongly consider providing 47 USC Section 230 protection to the Trust Bureau operator(s).

You can also encourage the private sector by clarifying the lawful structure of Trust Bureaus and defining their obligations. These must include user consent for all data use, bans on selling personal information, mandatory cybersecurity standards, and regular audits. You should also consider governance models that include consumer advocates, privacy specialists, and industry representation. This ensures balanced oversight and prevents abuse. If you take initiative now, you can help ensure that the Trust Bureau does not become a corrupted version of the system it is designed to replace.

Instead, it becomes a genuine public utility capable of verifying identities quickly and securely while exposing no raw data. The impact would be transformative. You would dismantle the monopolies of data brokers and background check companies that profit from consumers without empowering them and replace those companies with a system that is safer and more just.

You can then accelerate adoption by using smart incentives. Change does not always come from mandates. Sometimes it emerges from strategic motivation. You can introduce safe harbor protections for businesses that adopt Trust Credentials so they are shielded from certain liabilities if breaches occur despite their good faith efforts. You can strengthen penalties for organizations that refuse to modernize and then, for example, expose millions of SSNs in preventable breaches. To encourage consumers to adopt a secure, reusable, and portable Trust Credential, the federal government could make the cost of obtaining the credential a tax-deductible expense, similar to what the IRS did for certain parts of genetic testing via 23andMe.

Adjusting the liability framework will push boards to take trust seriously. You can also offer tax benefits or grants to smaller companies that implement privacy-enhancing technologies. Procurement standards are especially powerful. The US federal government is the largest buyer in the world. If you require vendors serving federal agencies to have trust-first verification and strong data protection, the industry will move rapidly to meet those standards. You can collaborate with state governments as well. Encourage states to integrate the Trust Bureau model into DMV systems, licensing, benefits, and other public services.

Your next responsibility is to strengthen consumer rights and raise the bar for corporate accountability. Give individuals the legal power to access, correct, and delete their personal data. This is not a radical ask. It is a core feature of the Trust Bureau model, where consumers are in control of their own information by default, not as an afterthought.

At the same time, raise the stakes for corporate negligence. Data breaches cannot continue to be treated as routine setbacks with minimal consequences. Update breach notification rules to ensure faster and more detailed disclosures. Impose penalties that actually deter misconduct instead of being absorbed as a cost of doing business. Companies should feel pressure to prevent breaches, not just to clean them up afterward.

Transparency will drive competition in the right direction. Businesses will begin to differentiate themselves based on how well they protect user data, and voters will support the effort. Identity theft, online scams, and digital fraud are not abstract policy issues. They directly harm people's financial well-being, their reputations, and their sense of safety. If you want to build momentum, bring real victims into the conversation. Their stories reveal the human cost of inaction and can help build urgency where technical arguments fall short.

You also need to look ahead. AI is changing the nature of fraud. It can blend stolen fragments of real identities into entirely new ones, creating synthetic profiles that pass traditional checks but wreak havoc downstream. AI is already being used to automate scams, impersonate voices, and manipulate communications at scale. The tools are only getting more advanced.

You cannot rely on outdated verification systems to keep up. For high-risk activities—such as opening financial accounts, applying for government benefits, or engaging in large transactions—you should consider requiring proof of personhood. This does not mean intrusive surveillance. It means designing systems that can verify a real human is behind the action, with consent and security built in.

This means ensuring that a verified human stands behind high-stakes transactions. Social media platforms may need to offer verification options so users can confirm that they are interacting with real people and not bot networks. For AI-generated content, you may need standards for labeling so citizens can distinguish truth from manufactured deception. You can support early research into digital trust marks for content and identity. Acting now will help prevent the next generation of threats from overwhelming the public.

Policymakers, history will judge this moment by what you choose to do. In previous technological shifts, such as the emergence of the internet and the rise of social media, law and policy lagged far behind. The cost of that delay has been enormous. Now you have the insight and the tools to course correct. You can embed the principles of data sovereignty and digital trust into the fabric of our laws.

You can create an environment where innovation thrives because privacy and trust are built into the foundation rather than bolted on as afterthoughts. You can also restore faith in public institutions. When citizens see leaders tackling the trust crisis with seriousness and compassion, they begin to believe again that government can serve the common good.

So move with urgency. Every month, more citizens are harmed by identity theft. More businesses face million-dollar breaches. The trust deficit widens. As a lawmaker, you hold the keys to change at scale, so use them to build structures that last beyond your tenure. In doing so, you will create a safer digital world for future generations.

Building Trust Together

No single group can repair the breakdown in trust on its own. Consumers cannot solve the problem without businesses changing their behavior. Businesses cannot make progress without policymakers updating the rules. And policymakers cannot succeed without consumers demanding better protections and transparency. Real, lasting change only happens when all three push in the same direction and hold each other accountable.

Privacy and data security are not niche issues. They affect every corner of society. When personal data is leaked, misused, or sold without consent, the consequences are real. It can mean identity theft, job discrimination, financial loss, or mental health strain. And yet, unlike other sectors, ordinary people don't have a powerful lobby fighting for them in Washington. There is no consumer data lobby standing shoulder to shoulder with lawmakers.

Farmers have a farm lobby, oil companies have an energy lobby, and banks have a coalition of trade groups that influence financial policy daily. But the average citizen navigating the digital world has no organized voice pushing for better rules. They are left to rely on slow-moving public pressure, scattered nonprofits, or lawsuits after the damage is already done.

This imbalance has allowed outdated policies to linger and corporate practices to drift into gray zones. It has also made it easy for lawmakers to overlook the human cost of weak data protections. That must change. Privacy is not a special interest issue. It is a public interest issue. And the sooner we treat it as such, the faster we can build a system that protects everyone—not just those with the loudest advocates.

Picture how this cycle builds. Consumers begin demanding safety and using Trust Credentials. Smart businesses respond by adopting them and showcasing their commitment to privacy. Policymakers notice voter pressure and industry movement and begin codifying standards that accelerate the adoption of Trust Credentials. As confidence rises, more consumers join, which strengthens the entire model.

HOW THE CYCLE BUILDS

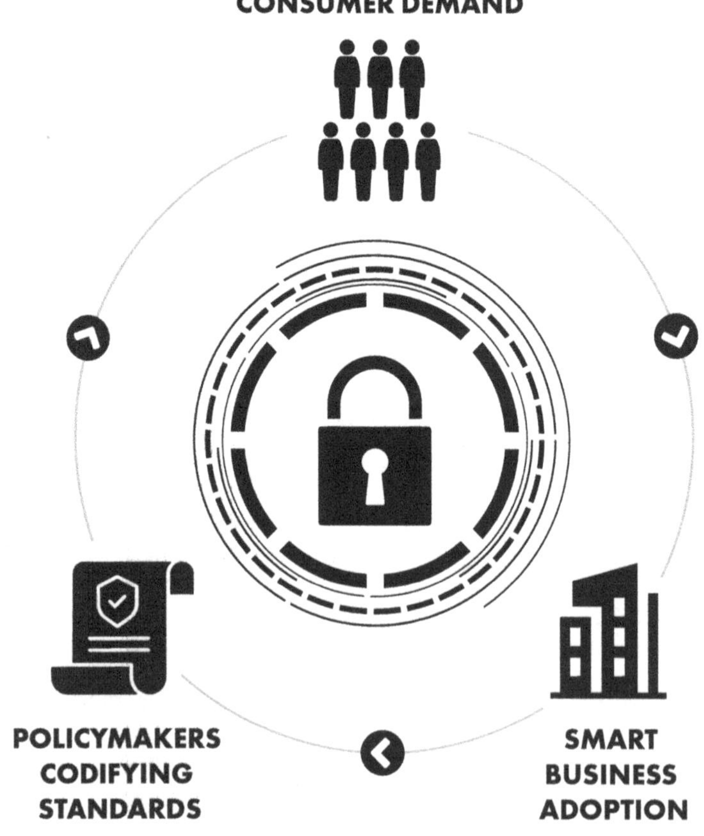

Imagine a reasonable timeline. In year one, the pioneers step forward. These are the early-adoption consumers, bold executives, and forward-looking policymakers who begin reshaping the system. Their progress may seem slow, but they lay crucial groundwork.

By year three, momentum reaches an inflection point. Major companies adopt Trust Credentials and even highlight them in national advertisements. More websites add a "Verify with Trust Credential" option, and users gravitate toward it because it feels safe and effortless. A few progressive states or countries integrate with the Trust Bureau for public services, proving the model works at scale.

By year five, Trust Credentials could be as common as driver's licenses for identity screening, or credit cards, or smartphones. Consumers will expect them as a basic right. They will wonder why anyone still hands over sensitive information when secure verification exists. Entire industries such as hiring, dating, rentals, and social platforms could shift toward trust-first systems because their leaders adopted early.

Now consider your position in this timeline. If you act in year one, you become a pioneer. You gain the early benefits of safety, influence, and participation in shaping the future. If you wait until year five, you become a laggard and face the highest costs. Laggard consumers endure years of preventable risk. Laggard businesses lose customers to competitors who built trust first. Laggard policymakers scramble to address crises that could have been avoided. The sooner each group acts, the better the outcome for everyone.

The sooner each group acts, the better the outcome for everyone.

This is the essence of the network effect. Every new Trust Credential increases the value of every other credential. Every company that adopts the system creates another safe space for transactions. Every new law that strengthens trust pushes millions of people closer to a safer digital life. Delay harms the collective, not just the individual.

To move forward, we must recognize our shared stake. We are building a new digital society where trust is woven into every interaction. Consumers demand better, businesses deliver better, and policymakers enforce better in a reinforcing cycle. The early scaffolding is already visible, with privacy expectations rising, companies acknowledging the link between trust and success, and legislators drafting new protections.

The key now is alignment. The Trust Credential and Trust Bureau framework offers a unified approach that brings privacy, safety, and convenience together. When we move in the same direction, the tipping point arrives quickly. Trust becomes the expectation rather than the exception.

If we delay, we risk creating a future where nothing can be trusted online. That would lead to fear, confusion, and economic stagnation. It could push people away from digital life entirely, forfeiting the benefits technology provides. That future is avoidable only if we act together and act now.

Collective action transforms what once seemed impossible into something natural. Companies discover they are not alone because customers reward responsible behavior. Policymakers find support rather than backlash. Consumers see that their demands matter. The effort becomes a movement.

The path forward is now unmistakable. The choice rests in our hands. We can drift toward deeper distrust or build a future where technology and trust stand side by side. Every era faces a defining challenge, and ours is to create safety and integrity in a world shaped by algorithms, data, and artificial intelligence.

Let us answer the call. Let us build this together. The choice is firmly in our hands, so let us choose privacy, let us choose action, let us choose verified trust, and let us begin now.

CONCLUSION

THE FUTURE WE CHOOSE

Throughout this book, we have confronted the uncomfortable truth that the digital world we built was rigged against us from the beginning. It asked for our personal data at every turn and then failed to protect it, leaving us exposed to risks we did not consent to.

We examined how this broken system harms real people. We saw individuals defrauded by scammers, families devastated by identity theft, and countless opportunities lost because no one could trust what they saw online. These were not abstract problems. They were vivid reminders of a growing crisis in human trust.

Yet that was only the start of the hero's journey. In the face of darkness, we searched for the light, and we found it. We uncovered the idea of the Trust Credential for Life, a transformative way for individuals to carry a secure and verifiable identity and background proof they control fully. This was not a tweak to the current system but a complete rethinking of how identity and trust should work.

We then explored the Trust Bureau, a bold redesign of the infrastructure required to make this vision real at scale. It places consumers at the center and enables identity verification without exposing raw personal data again and again. These tools became the elixir of this journey. You learned how they work, why they are different, and why they can succeed where previous approaches failed.

Most importantly, you realized this solution is not science fiction. It is already emerging in real-world practices such as continuous screening, tokenization, and user-controlled consent models. What once felt out of reach is now within sight. The path to a safer and more trustworthy digital world is no longer theoretical.

Now, as in every great journey, the final and most important step awaits. This is the return home, where the hero carries the solution back to the community and uses it to heal and rebuild. The journey does not end here. In truth, it begins here. The knowledge you have gained must now be put to work in your own life, your workplace, and your sphere of influence.

To understand just how high the stakes are, we must look at the two possible futures ahead. Think of them as Future A and Future B. One future arrives if we fail to act, and the other becomes real if we commit to the work. The difference between them is nothing more and nothing less than our collective will and our collective action in the years ahead.

Future A: A Bleak Future

Imagine the year 2030 and a world drowning in a deep trust crisis. Data breaches have become so common that they barely make headlines anymore. Hospitals one week, a bank the next, a government agency right after. People feel numb to the news, yet the damage keeps piling up beneath the surface.

Almost everyone has been affected. Identities are compromised so often that many households juggle multiple credit monitoring subscriptions and constant fraud alerts. Even with all these tools, the attacks keep coming. AI-generated synthetic identities spread unchecked and fuel fraud at levels the world has never experienced before.

Communication channels are flooded with sophisticated bots and deepfakes. Phishing emails arrive with perfect imitations of familiar voices, created by AI that can trick even the most cautious individuals. The line between real and fake blurs until the average person no longer knows whom to trust. Every message becomes suspect.

In this environment, society grows increasingly guarded. People hesitate to shop online or share any information unless absolutely necessary. Innovation slows as consumers and businesses alike move with hesitation, weighed down by the constant fear of being exploited. Trust, which once fueled digital growth, dries up and leaves behind a brittle, stagnant economy.

Legitimate users endure endless hurdles just to prove they are real. They juggle multiple passwords, CAPTCHAs, and identity quizzes asking for childhood details. Meanwhile, bad actors still slip through cracks by exploiting outdated systems that cannot keep up. Honest people struggle, while criminals adapt and thrive.

Fear becomes the default state online. Social platforms that once buzzed with activity become empty shells filled with bots and burner accounts. Real people withdraw from public spaces or lurk silently in the background. No one feels safe enough to engage, because no one feels certain who or what is on the other side of the screen.

As trust collapses, darker forces move in. Governments facing chaos reach for heavy-handed surveillance measures in an attempt to root out fraud and crime in the internet commerce. These desperate actions erode personal freedoms and tighten restrictions on everyday life. The individual becomes smaller as the machinery of control grows larger.

Personal sovereignty disappears. People feel as though their most private details have escaped into the world and now live in countless corporate and criminal databases. The fictional idea of a data dictatorship has become reality. It is a world where privacy is shattered, autonomy is gone, and yet security remains out of reach.

The most painful part is how the damage falls hardest on the vulnerable. Elderly citizens lose their retirement savings to convincing impostors. Job seekers face rejection because of identity mix-ups or false records they cannot correct. Businesses collapse under the weight of constant fraud and charge-backs, as insurers have stopped insuring them for data breaches and cyber threats because of the sheer volume of occurrences. No one escapes untouched.

The insurance industry saw it coming first. Over the years, providers began withdrawing from high-risk regions where claims from natural disasters—wildfires, hurricanes, earthquakes—kept rising. Eventually, they stopped writing policies for properties in areas with repeated losses.

The same pattern is unfolding in the digital world. As cyberattacks grow more frequent and damaging, insurers are reevaluating their willingness to cover organizations that store large volumes of unprotected personal data. Payouts are rising. Premiums are climbing. Coverage limits are tightening. In time, companies that continue to operate without strong data safeguards will be priced out of coverage or dropped altogether.

It is a bleak, exhausting reality that touches every corner of society. And it is entirely plausible if we continue down the current path without intervention. This nightmare is not a distant fantasy. It is the logical end point of inaction, complacency, and systems that were never built to withstand the world that now surrounds them.

Future B: A Bright Future

Now imagine an alternate 2030. Data breaches have sharply declined. Hackers still exist, but they no longer have easy access to massive stores of personal information. Organizations no longer hold sensitive data they do not need. Instead, individuals carry their own verified information in a secure Trust Credential wallet. Identity and eligibility are confirmed through cryptographic proofs that reveal only what is required for a specific interaction.

The change was driven by necessity, as the financial risk of maintaining outdated data practices became unsustainable. Businesses that shift early to credential-based systems will be better positioned. They will hold less sensitive data, reduce their exposure, and become more insurable. Consumers will benefit from fewer breaches and more control. And insurers will have a clearer view of who is managing risk responsibly.

When attackers try old tactics, they find nothing of value. Databases reveal only useless tokens that cannot be reverse engineered. Breaches still happen because no system is perfect, yet the damage is contained and insignificant. Cybercriminal networks begin to dissolve because the financial reward disappears. Trust in online services climbs as people witness that their information is finally safe.

The digital economy enters a period of rapid growth fueled by confidence instead of fear. New services emerge in areas once viewed as too risky. Peer-to-peer lending becomes mainstream because trust profiles allow individuals to evaluate one another instantly.

Community sharing platforms flourish for the same reason. By 2030, Trust Credentials are everywhere and woven into daily life.

You might have yours secured on your device and linked to a biometric identifier, so it is undeniably yours. You use it to log in without passwords, to prove your age at an online store with a single tap, and to pass through airport security as easily as travelers with TSA PreCheck. Imagine not uploading ID scans or exposing private documents ever again. Safety and convenience finally move in the same direction.

Life becomes easier and safer at the same time. Social networks and dating apps become far less chaotic because verified human profiles become the default. Trolling, bot propaganda, and predatory behavior decline sharply. When a message or a piece of content reaches you, you can tell whether it is genuine because it carries a small trust indicator. Misinformation loses its power because verified truth carries more weight.

The economy benefits as companies save the massive amounts of money once spent on dealing with fraud losses, build out identity screening systems and infrastructure for their own purpose, litigation risks, and compliance burdens. Those resources are redirected into better products and greater innovation. Individuals who were previously underserved by traditional systems now gain access to opportunities that were once closed to them. The single mother denied housing because identity theft damaged her record now carries a Trust Credential that proves her integrity and clears her name.

The gig worker who used to dread repeated background checks simply shares their trust profile and is hired on the spot. Barriers fall away because identity and reputation become tools people control rather than obstacles thrown in their path. Digital sovereignty becomes real. You decide who learns what about you and how much you choose to reveal. Your ability to present your truest self is finally in your hands.

The Movement We Build Next

These two futures diverge sharply, and the fork in the road is directly in front of us. The next few years will determine which direction we take. Future A or Future B depends entirely on what we do now. The difference between a society that breaks and a society that rebuilds comes down to the choices we make together.

So I invite you to stay connected, not just with these ideas, but with others who share this mission. The pages of this book may end here, but the work continues. So share your stories. Tell me what challenges you are facing and what solutions you are testing. Parents protecting their children. CTOs building trust-based employee credentials. State officials rethinking digital identity. These stories help fuel momentum.

As we connect, we can learn from one another and accelerate the movement. I challenge you to recommend this book to someone who needs to read it. Quote it online. Bring it into your workplace, your classroom, your boardroom. Every conversation makes Future B more real and more achievable.

It's my hope that in the next five to ten years, every person in America has a digital Trust Credential in their wallet, ready to use across jobs, services, platforms, and institutions. Just as a driver's license from one state is accepted across the country, your credential should travel with you and be recognized wherever you go. America can be the first country to adopt this revolutionary approach and be the leader of the AI-driven digital world.

This future is within reach. Verifying your identity securely should be as easy as saying, "I'll Trua it." That is the end goal: to embed trust into our daily lives so completely that it becomes second nature.

Raising awareness is part of the work. We need public messaging that's direct, urgent, and hard to ignore. Picture a Super Bowl commercial that simply says, "Stop giving away your Social Security number." Or a national campaign that asks, "Who owns your identity?" These are the messages that spark conversation, shift behavior, and change culture.

You picked up this book for a reason. Something in you cares deeply about the well-being of your family, your community, or your society. That tells me you are already the kind of person who can help move us toward Future B. Let us choose the path that restores control. Let us build a society where trust is measurable, mutual, and lasting.

We now hold the tools. What remains is the will to use them. The future is not fixed. It is shaped by what we do now. Let's walk this road together. Let's build momentum. And let's reclaim trust and demand how it is verified—one action, one decision, one connection at a time.

INDEX

C

D

E

O

P

Q

R

S

T

U

V

Z